<u>Divine Doorways & Personal Growth</u>
<u>The machinery of God & spiritual doorways</u>
<u>Texting Heaven: R U U©</u>
<u>A Writer seeks Truth—a spiritual diary.©</u>

Dana Fitzgerald

Our limited understanding of dimensions beyond what we sense leave us with many questions and mysteries: what follows are notes on life, Biblical questions, the mystery of history, UFO tales, ET, the universe, Velcro, Mr. Clean, and Gummy Bears. And, what's up with Roswell?
This is a daily quest by the author of: *Drugstore Cowboys*. Throughout history, ignorance of the unknown created fear of the unknown; with open eyes and open minds, we should accept the fact that we are merely scratching the surface of what lies beyond our world, and even the forces our world generates are beyond what we know. Gravitational and magnetic anomalies, crop circles, an entire historical record of unknown building and mind-boggling construction techniques should tell us to be ready for anything. Unfortunately, our arrogance and smug understanding creates a barrier between what could be possible, and what is considered impossible. For many years, sarcastic egotists claimed if man were meant to fly, he would have been born with wings. We all know how that turned out. Lessons like that should make scientists ready to accept as reality what is now seen as magic; unbelievable forces are at work, and our great geniuses are either covering the truth, or too stupid to embrace forces beyond our current abilities. Anti-magnetic devices, inter-dimensional doorways, UFO sightings, ancient rocks cut with laser-like precision, are but a few facts we observe, but cannot explain. The basis of any science is to observe a phenomenon, accept various observations, and work towards a unified explanation. Doorways to the divine, pathways to other dimensions, and incredible power harnessed with ease are observable facts that we should

be studying instead of shunning. The early Greek inventors and philosophers were able to achieve miracles of science because they weren't limited with a set of physical laws, or tables of elements that are considered unchangeable properties of an incredible universe.

This is exactly what society needs; *another* book. Then again, this is a book that opens minds, and opens spiritual awareness, necessary components to understand an amazing amount of information we humans have not yet acquired. Past attitudes towards discoveries that are considered science fiction are alive and still pulling a hood of ignorance over our eyes…what we need know is imagination to join science and uncover the mechanics behind effects that well beyond our level of technology.

With caution tossed out the literary window, we're on an alarming and injudicious jaunt through the psychogenic interdimensional generator of my brain. This mélange of mischievous mumblings takes us through the mind of a mythological schizoid, as selected words flow from my dithering digits or fickle fingertips. Touching on life, half-life, no life, historical histrionics, current conundrums, Biblical benefits, story starts, side-trips, and denouements, ideo-ideas, and farkleberry frappachinos—this is a flow of consciousness diary that peeks inside the insane imaginings of my mind. Describing a wistful, woozy wonderland, explored in a wizard's worthless Winnebago, this imaginary realm is surrounded by a tract of imaginary but devastated post-apocalyptic wasteland: it examines what is true, how we know it is true, and what to do with this truth. Truth is seldom simple, and is argued *ad naseum*. This is a journal that some sneaky, underhanded people secretly opened to read, and all agreed it would make an entertaining endeavor. You'll never get sick of the story line…it changes each day, from one style to another. This mental mix is a hard-copy video clip of creative transference…I think, and voila, words appear. These questions are timeless; conundrums of antiquity. Like the Roswell rock: a very hard, andesite stone, with a complex geometrical pattern carved with laser-precision on its surface, presents an interesting question, and the fact that this particular piece of common andesite is also magnetic create another question with no answer. On top of that, an airplane flew over a field, landed, and flew over the same field ten minutes later and discovered a crop circle that is identical to the debossed image on the Roswell rock. Strange? Undoubtedly. This is something that is not within our ability to scientifically explain, and therefore creates derision and scorn from frustrated scientists that can do amazing math, but cannot explain these two observed and recorded facts. There are so many, I could fill this book with equally amazing facts that science cannot answer…their only explanations are to yell hoax, or just shove it under the rug. It's been suggested there are more artifacts hidden in museum basements than there are on display. Why are we so afraid of what we don't understand? The early church used to cry heretic and burn anyone that questioned phenomenon at the stake, and then hid the evidence in an ever-expanding collection of misunderstood writings, puzzling artifacts, and attempts by philosophers to answer these questions that are still unsolved. Governments refuse to acknowledge certain events because they claim the world is not ready for the answers—I think we are ready for the answers, and have the right to make up our own minds whether to believe or ignore. We are allowed to make up our minds on one of the greatest questions to ever change the human race—whether the Bible is truth, and that answer opens spiritual doorways to heaven and hell, angels and demons, and life after death. If we are able to make that decision, why would a UFO make any difference in our world-view, and why are we not allowed to know what they know. The men-in-black are real, and they cover-up all evidence of these events. There was a television show that only lasted 6 episodes before the government started to get worried: Hanger one-UFO case files. One could only surmise the show offered too much evidence and asked too many questions the government has firmly decided are beyond what the world of

2014 should be allowed to ask or deserve a real answer. This book wavers over the line of what we are permitted to know, and what we are not permitted to know, and asks why.

Remember, questions are harmless; answers, or truth, are what create panic. When the questions pile up, and are asked by everyone, governments don't provide the answers, they hush up and end the questions. That's another statement that can be backed up by a mountain of examples—this isn't a list of questions, it provides a list of possible answers, and each reader must decide what to believe. If I had proof, I'd be on CNN: consequently, I'm offering solutions that are within our ability to understand. It takes faith to accept certain answers, and faith is the basis of every religion. Jesus said, "Blessed are they who believe and have not seen." Sometimes, all you have to go on is logic, guesswork, and a whole lot of faith. The Christian religion appeared in a time of magic, mysticism, and a lot of unknowns. When people saw loved ones raised from the dead, they believed; Jesus himself appeared to over 500 people after his crucifixion. It was a simple time, and people tended to believe what they saw; nowadays, experts toss around theories, possibilities, and explanations that keep skepticism alive and well. Scientists don't operate on belief and faith—they want empirical facts that can be proven by the accepted scientific method that sometimes obfuscates the truth because there is no known test to discover an unknown event or creation. Sometimes, you have to have faith; we have faith in certain people, so why not extend it to encompass and explain what we know happened, but don't understand.

Answers involve leaps of faith, open minds, and a willingness to look underneath that ever-present blanket of science. The one that covers anything it can't explain. This literary road contains potholes of dramatic, but descriptive conglomerations of verbal conundrums, alliterative alligators, and other *jeu de mots* that spark the narrative, and add that sizzle you hear when backyard barbeques turn otherwise good cuts of meat into burnt offerings even the dog ignores. This shows that our reality is covered by an invisible wall of other-dimensional worlds, real places that sometimes intersect our standard point of view. Electro-magnetic forces open doorways into these realities, and some step through without leaving a trace...the Bermuda Triangle phenomenon, or the Philadelphia effect is a real and dangerous portal visitors seldom return from. Government whitewashes aside, we need to accept these truths, and see our world as magical and full of unknowns. Ignorance of the unknown produces fear of the unknown; once it is scientifically explained, we chide ourselves for being childish, and move on, our realities intact and our minds still in one piece.

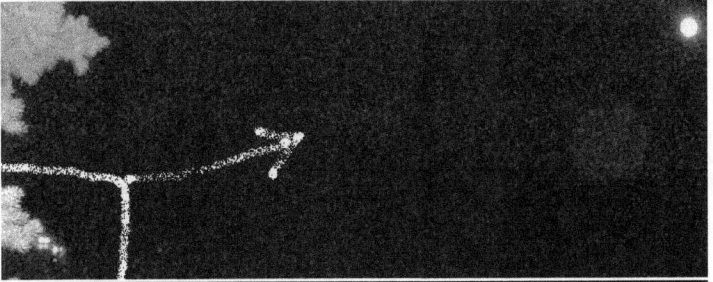

Notice the bottom lights in the middle of the upper tree, and the moon. My camera got this moon shot, and the lights are at the top of a 300 foot tree.

Dear God: & all on a higher-plane of existence (anywhere but Earth)

As the author of the real-life book ***Drugstore Cowboys***, I've enjoyed a rather bumpy road in life, and still find bumps as I slog along. I found love was like a special Sunday dinner, only given when I did something my family approved of, and something they could brag about to the rest of the family. When examining life, I find the reassurance and salvation of the historically proven Bible uplifting and a source of joy in an otherwise dreary life. It has even more appeal today than it did back in the day, and back then, it was magical music to an enslaved, deprived, and beaten society, who often live hand to mouth day in and day out. I like to think there is a meaning to existence, and the Bible teaches exactly what I need to hear. It may be written in a mythical manner so primitive societies could understand its moral lessons, but that should never detract from the wisdom it contains. It is the closest thing we have for an operating manual for spaceship Earth, and should be seen for what it is; a book that puts the fear of God and polite behaviour into an early society that only knew war, bloodshed, and the power of the sword.

With so much mystery to life, it's refreshing to hear about salvation, a source of redemption that begins with the creation of humanity. Whether it was DNA added to Neanderthals or evolution—it doesn't matter. Optimism over pessimism, good over evil; light will always eat away darkness, and an eternal source of light is the essence of God.

I find it hard to truly accept forgiveness, a personal problem of mine, but the Bible clearly states that if you forgive, you will be forgiven. The rest is a matter of faith. I'm not some evangelist, and this is the first time I've ever talked about my travels with God, and my rough ride to redemption. It takes a lot of modesty, faith, and prayer. I'm a little low on trust, so finding a church that doesn't seem like a pure scam to get collection dollars is a long one. I'm not putting down church, or the idea of church, but I find a big corporate style church has too much of a business sense to it, and it tickles my trust meter. A church is supposed to be a true body of believers who work together for the glory of God, and serve others before serving yourself. Christ emphasized that He came to serve and save others, the real, altruistic practice of Christianity, and something you have to search for, as many Christian Corporations sadly put the bottom line before the needs of their congregation. A true church is not a time or place, but a chance for believers to unite and help the less fortunate and the suffering. We are to help orphans and widows. Alas, the hucksters of old are alive and flourishing, selling God's word, and using his name to make a few bucks. Tammy Faye comes to mind, (with nightmarish visions of too much mascara) as do many other *televangelists*. Look at Jim Jones and Kool-aid…wow.

I'm sharing my quest for God and spirituality here, and who better to trust, than someone with a long history of drug addiction and sales…someone that really needs forgiveness. If I can be saved, anyone can be saved. There's some short stories tossed in, so it's an on-going reading smorgasbord. I can promise you won't be bored with a lot of church talk and Biblical quotes. I don't want to repeat that, but I do want to emphasize this: I have no church sponsorship. I'm just a regular guy…I don't have PhD's in Theology or advanced training in Christian Orthodoxy—I'm a historian, but I only have an Honor's B.A. in English and History. That's like having slightly more than high school these days; we have a glut of over-educated, under-utilized graduates, and I can't afford to keep up, and still owe on a lousy Computer diploma…a scam from the diploma mills. If I kept at it, I'd be in so much debt, they'd garnish my old age pension…don't laugh, they still might. Too many fell prey to these Colleges, and they still churn out graduates, advertising the latest "flavor" of the year in careers. What a waste.

This is a narrative that records my search for the true meaning of Jesus, God, Spirituality, and cosmic creatures, like Alien and ET. Believe me, they are real; I have a picture of one taken from 30 feet, and there are so many video clips, pictures, and eye witness accounts around, they can't keep denying they exist forever…soon, someone will find something that they can't refute, and floodgates will pour out the hidden truth, ever since the Roswell crash turned into a weather balloon…or was it swamp gas. Who cares, everyone knows the government is lying, so why they even bother with stupid cover stories can only mean there's a lot going on, and they are afraid of what that knowledge will do to an already messed up world. Did they ever consider that proof of UFO and extraterrestrials might actually help our world unite? Whatever, they love their secrets, but they aren't fooling anyone…nudge nudge, wink wink.

Anyway, focusing on the world, charities, good deeds, things to do, attitudes, science, and even Velcro, everyday events or urban myths are talked about in an analytic and humorous manner. Would you believe me more if I said I had a picture? I've recorded my personal discovery of what Jesus means to me, and explain why this is a very logical postulation; if someone reads the Bible carefully and doesn't get swept away by convoluted interpretations our "Preachers" endlessly spout—everyone can see the truth in the words, for even Jesus said we are to seek our own enlightenment, and not follow self-proclaimed leaders blindly. Yes Tommy, don't believe everything you heard in Sunday school. It's possible to mix true wisdom from other religions, as long as you bear in mind what really matters. There is a lot of easy to understand mythologies, simplified moral lessons that would help a bloodthirsty and savage population, something that's hard to do when the entire world lived and died by the sword…or famine.

God: such a short word, yet containing such immense power…abilities beyond our understanding—supremacy outside our perception. Throughout history, the notion of God has explained what humanity saw as magical, and overcame their limited outlook on the world. It was a time when people worshiped Idols, a pantheon of different Gods, and pure evil; they also worships ET's that showed up in amazing machines with technology that looked like pure magic, but even some of our standard technology would be viewed as magic to these pre-industrialist and primitive civilizations. In literature, impossible situations and inexplicable positions are saved by "Dues ex machina," the God from the machine; whenever a hero has no way out, this plot device manages to save everyone, and hence the solution is known as a "God-like" intervention—it is believable, for no one doubts the power of God. Our technologically arrogant society might dispute the existence of God, but everyone accepts Godly power—unfortunately, too much science fiction has turned God into an alien, or some all-knowing life-form that exists in a higher dimension. Science, evolution, and advanced medical knowledge uncovered some of God's secrets, but instead of being overwhelmed and amazed at the complexity of creation, some hard-nosed technocrats chose to believe everything has a scientific explanation, and like maturing children, they believe they've outgrown the need for God. Rubbish, I say. The miracle of life, the mystery of the soul, and the proof we have in the Bible is enough for me to believe what Jesus came to share—the chance of redemption through faith and forgiveness. It provides many arguments, and I will endeavor to share what I've discovered. Eternity is a long time, and making the decision to trust in an all-powerful God shouldn't be taken lightly. Living a Godly life is a personal quest, and cutting through the many obfuscating opinions about redemption requires nothing more than an open mind, and a contrite heart. If you're too arrogant and puffed up with self-importance, this might be something that seem impossible, but nothing is impossible with God. Just keep an open mind for now, please?

Something that really bugs me is the tendency for people to ignore the Bible, and put their own "spin" on what it says. Twisting the word of God to suit personal likes and dislikes won't get you into Heaven...humility and repentance are required, and are not that onerous or hard to live with. There is nothing new under the sun, and every worldly explanation for Jesus has been thought of and thrown out there as a possible theory. Truthfully, they are just another way to avoid living a just life, letting people feel comfortable in our material world, and avoiding the guidance of the Bible. Living a just life benefits everyone, and is necessary to embrace the love of Jesus and gain eternal life. Suggesting Jesus was a "master" in some secret knowledge, some old science, or even having traveled to India to study meditation and methods to slow his heart down to pull off His "resurrection." They take His powerful words, but miss the really big point. He was the Son of God; he wasn't some special charlatan, mystic, mountebank, or side-show huckster. He was who He said He was, and the miracles He performed, like raising someone from the dead, could only be done with the power of God. Many theorists prattle on about lost knowledge, past super civilizations that were more advanced in the "science" of the soul, or extraterrestrial technology that was lost.

Of course, the massive platforms at Baalbek, Sacsayhuanan, and other megalithic sites suggest our ancestors knew a lot more than we give them credit for. The Baalbek stones are doubly mystifying, as the massive, 1,500 ton stones are on top of a foundation and not the foundation. Why were these colossal stones needed? A UFO landing Pad...perhaps, and just a guess. The Sacsayhuanan stones are the most precisely carved examples of Andesite and Diorite in the world. Extremely hard stone, it is mystifying to imagine how they were made: some suggest alien technology and Lasers, others postulate incredibly advanced methods that soften the stone so it could be scooped out. We see the relics, and use modern knowledge to surmise how they were made;

from the distant past, these remnants baffle us, and defy traditional manufacturing techniques from the old world. They are enigmas of rock; unable to be dated, we only know their creation involved methods we have yet to discover or understand. Mind-boggling would be an apt description, and they will remain a "mystery from history," until more is known.

Trying to extrapolate all their different knowledge into "cycles of time," using the many exact calendars they left us (the Zodiac and the Mayan Long Count, e.g.), is merely modern pseudo-science, or new age thinking. UFO phenomenon have been with us a long time; what their agenda really consists of is something we can only speculate on, but I would trust in the power of God any day, rather than trust some little Grey alien with a fancy flying saucer. Abductees haven't shared precise plans with us, they only describe horrific medical experiments; naturally, something that any curious, scientific society would do to an unknown specimen. Pinning your hopes on them is like hitching your wagon to a star—you never know where it will go, and you don't know what to expect. Many authors debate this, provide their own personal beliefs, and the whole "spirituality" question gets muddled, mixed up, and moved away from where it began. The Bible; the only real operating manual for spaceship Earth, and the only guide to a righteous soul…you don't need to take everything in the Bible literally, for it was written for any primitive society, but the lessons on morality and pure righteousness are spot on, and thanks to the Divine Intelligence that created our world. Why is it so hard to call Him God, believe He shared His words through prophets and through His Son? Usually, it's as simple as this: they don't want to obey the suggestion to ignore material possessions, and are too dazzled by the splendor of this material world. Jesus said that after death, Heaven would amaze us with its splendor, and the only light needed is the light of God. Why is that so hard to accept? Is it simply because they want the power of the Bible, the "blessings" and the redemption, but don't want to be a Christian, go to Church, and announce to all that Jesus died for our sins? That "old-time" religion is misplaced in our "new-age" society; since when did humanity suddenly adopt airs? The answer is in the last century; our technology grew by leaps and bounds, yet our spiritual awareness is still unresolved

They have to wrap it all up in modern techno-spirituality, some sort of science of the soul, a lost knowledge known to the Druids, Atlantis survivors, Egyptians, or ancient civilization that have only left us megalithic mysteries our historians pooh-pooh and ignore. Don't they recall how the God of Israel beat Egyptian magicians, and allowed the Exodus to happen? The uncanny attributes of the Ark of the Covenant, instead of being accepted as a gift from God, is now seen as some sort of alien technology, something that fits into our little superior scientific outlook on life. They have their nice neat timeline, and understanding Gobekli Tepe, or other discoveries, threatens to blow it away with new discoveries. The obelisks at Gobekli Tepe are suggested to be 14,000 years old; man was coming out of the ice age then, and supposedly didn't know how to carve monoliths, or organize enough to erect massive circles of them. Strange, but absolutely true. Or the Bosnian pyramids; one apparently still shooting out a microwave at 33 KHz...how or why? Even the cement used was superior to modern cement, as it was a type of polymer, something previously undiscovered. It seems the more we dig, the further the gap in what we demand must have happened, a scientifically explainable historical progression...but what really happened, all those shrouded centuries ago, now covered in the mists of time. Leftover evidence is everywhere: the Nazca line, the Ica stones, Lai lines, precisely constructed cities, all on a circular path that suggests they were in contact, and built them in full knowledge that they were building them in a measurable circle, and all on specific geographical locations with similar properties. Only God could have wiped them out. Early man had access to sciences that have been lost; projecting magical powers to them doesn't explain everything, but is a cool thing to think about, and a wild source of speculation that fits into the new world view of aliens, super technology, and lost knowledge. Some of that knowledge could be hidden by the Vatican, as it was the only book depository after the library of Alexandria burned. Sorry, I don't trust the Catholic Church...I've read too much history, and know all the horrendous acts they committed in the name of God and human greed. The 325 Council of Nicaea gave the Church a lot of power, and sadly mixed man's greedy nature with the righteous teachings of Jesus.

Buddhists teach personal awareness, a state of existence that is good; likewise, many teachings share the idea of spiritual awareness, and show there is a real difference between good and evil. Being spiritually good is a universal emotion, a feeling that transcends existence and is therefore beneficial to all; add to that and you should be in good shape, but you must make your own choice. I don't profess to know all the answers, but I have worked out a few things that I believe helps my spiritual sanctuary. I also recognize I need help, which is sometimes a major difference between righteousness and arrogance. Spiritual arrogance is not the door to enlightenment, Nirvana, or Transcendence. People often focus on enjoying life, but pursuing a healthy spiritual center is paramount to an eternal future of love. Only the love of Jesus can quench your spiritual thirst, as he said he would give "rivers of living water" to slake our dry throats.

Enlightenment is a state of being, Nirvana a plane of existence, and walking and talking with God means you're in touch with the Divine (or slightly Schizoid). Transcending our lower, feral, carnal nature, and becoming a creature of love and light is something everyone should desire and try to accomplish. Mastery of one's life gives material benefits, and spiritual awareness is something you must attain before you really understand what benefits you acquire. People devote their entire life to this, and try to "ascend" to a higher state of consciousness. Some get quite crazy, as people commit mass suicide, just because some shifty spiritualist conned them into believing what they preach. LSD might help…there was Charles Manson, Jim Jones, with the Jonestown mass suicide, and many more, like the tragedy in Africa—sadly, we only find these cults after they commit a final act, like mass suicide. Jones was so dangerous, they should have sent a Delta squad instead of a fact-finding Senator—they were murdered, an act that will send him to hell…I hope.

Heaven's Gate, formed by Marshall Applewhite, believed an alien space ship following the Comet Hale-Bopp would take their souls into the pristine state of being, somewhere out in the vast universe. I think 38 people fell victim to this nut-job; taking your own life for something you believe falls into free choice, but dragging a bunch of victims with you borders on Hitler-style evil. Some people are gullible and easily fooled. I don't know how these con artists brainwash so many people, but if drugs are involved, anything goes. Feed you followers LSD, along with high doses of friendly opiates, and people might believe anything.

In the 1979 book '**Messengers of Deception**,' Jacques Vallee looked into the background and belief system of this mass murderer—he discovered a well-populated psychotic world, its members easily convinced absolutely insane predictions were probable. The book was ignored for two decades, except for UFO guys, until the final act shocked the world. The movie "Mysterious Two" was loosely based on Applewhite, due to attention in the 70's. Heck, for all we really know, perhaps he accomplished his "mission." What's weird is they all carried a five dollar bill and three quarters for interplanetary fare. Wow.

Like any successful mad bungee-jumper, he had a very well reasoned belief that included the Bible, a near-death experience, and science fiction. Marshall believed he was directly related to Jesus, and believes he was in some "Evolutionary Kingdom that was on a level about mere humanity." When someone likens themselves to Jesus, that's usually a huge red-flag shouting "psycho." The Bible specifically cautions us about "psuedo-Christ-like figures" that pretend to be Jesus, and mentions that avoiding these guys with a ten-foot pole is a good idea. That's common sense, and specific advice from all the apostles, so I'd tend to believe the people who actually knew Jesus, rather than some modern-day extra-crispy that claims the sky is falling. That's exactly what Applewhite thought: he said Earth was about to be "recycled," and the only chance to survive was to leave…immediately, minus your body.

Similar to Ford Prefect in *"**The Hitch Hikers Guide to the Galaxy***," he tried to hop aboard a passing UFO, but he didn't have that magical "Hiker's ring" to catch the Vogon ship. He should have read the book first, and realized a life with a bunch of nasty and fat accountants, only concerned with tea time and lunch, would be worse than life here. Taking your own life is one thing, but taking others with you doesn't compute. They referred to our bodies as vehicles for our souls—in an interview, Rio DiAngelo commented on a picture of his son, saying, "Look, there's the little vehicle." Rio was one of the only survivors…supposedly left behind to explain things. The whole thing sounds so insane it begins to take on some logic of its own. A very small piece of logic, as killing yourself so your spirit can hop on a passing UFO is really going out on a limb. I suppose we'd all like to believe in something that would give us instant meaning, so "projecting" an insane notion over your life gives your life extra meaning. Instead of Joe-Schmoo, who works as the night manager at Taco-Buritosaurus, he becomes some super evolved spiritual life-force that transcends the Earth, and leaves the less informed behind. A humdrum existence is propelled into a galactic adventure. Heaven's Gate disciples were uniformly dressed: identical clothing, brand-new Nike running shoes, and a purple square draped across their chests. Their armbands read: "Heaven's Gate Away Team." Among the dead was Star Trek's Uhura's brother, Thomas Nichols, the brother of actress Nichelle Nichols, another Star Trek link.

Borrowing truths from the world's different religions, many work out their own "master plan," without using the leavening guide of common sense. It's true that faith is required in attaining eternal life, but faith in Jesus, not in some UFO that may or may not exist. For over 3,000 years, the guidance of the Bible has been enough to satisfy most people—it's only the modern-day spiritualists that create a mish-mash of beliefs, diced, sliced, and blended together to make some super interesting outlook. Heaven's Gate professed shedding material loves: family, friends, sexuality, individuality, jobs, money, and all possessions. That's almost right out of the Bible, as we all know we can't take anything with us when we die…the only thing that matters is what you believe, and the good deeds you've done, or bad deeds, to keep it balanced.

It's like the old Bible is too mundane and boring: it's available to everyone, so they have to create a special religion to stand out from the crowd. They look at the new knowledge we have, and then pick and chose what their new religion should have. One thing that's always consistent is that only "a select" portion of the population can continue, and only the "chosen" are allowed in these groups. Only the crazy would join these groups, so that makes sense; telling them they are "chosen" plays to their ego, and completes the whole charade. Con-man tricks 101; only con the con-able—believers who are easy prey. Some take things at face value, others want a more "realistic" assessment, and are the sort con men avoid. Only scam the scammable; this is a truism that permeates all these special groups, from Jonestown to that village in Africa, people buy into what they want to believe…so much so that they are willing to die for something that resonates within and provides that sense of being different or other worldly.

Hordes dislike the way of the world…the inequality, the injustice, the overt frauds many governments perpetrate on their citizens. They are elaborate versions of con-men, but benefit more people at the top; greed pushes men into power, and sometimes the love of power guides them. When we all learn to see this, it's like our eyes are opened to the ways of the world; the so-called "great" men are shown to be no more than greedy, power-hungry, infamous, and driven by self-absorption, self-confidence, and self-aggrandizement. There are altruist humanitarians that truly try and help the world through charity, but they're usually rich to begin with, and really do have "spare change." Anyway, UFO religions seem cool and "modern," so they'll always have many applicants. I remember reading Heaven's Gate purchased "alien abduction insurance," a purchase that covered 50 members and cost $10,000…wow, they'll insure anything. I suppose they didn't want to hop on the wrong ship. I would say something about Scientology, but I'm afraid; they really do have the power to give you a very bad day.

A very thin line holds my inner madman at bay. It gives me story ideas, prose, and the odd output of random synaptic pathways. Odd would be a generous conclusion, as I question everything from UFO's, ET, and history; toss in everyday life and human interactions, and it becomes a wacky memorandum that stretches my mental security on a daily basis.

An honest buck is hard to make; the margins are low—too many people are doing it. Evil prevails when good men fail to act—that saying spins both ways, and sadly shows that the two are constant, and need each other to exist.

Day 1:

A short history of change: Hard line Orthodoxy is questioned.

Science fiction is a wonderful flight of fancy, but no matter what story I've read, they don't include the age-old wisdom of the Bible. Some have an updated, more advanced comprehension of a "universal spirituality," an attitude many see as common sense, something befitting an enlightened society. As the Bible states, "They perceive but never see, hear but never understand…for this people's heart has grown dull, and with their ears, they can barely hear, and their eyes they have closed…lest they should see with their eyes and hear with their ears, and I would turn and save them." (Matt 13: 15, Isa 6:10) Sadly, many fail to remember, read, or understand some very important paradoxes:: the Bible was a time machine, and the prophets of old accurately predicted the arrival of Jesus, the deeds He would perform, and even quote the words He would say.

More accurate than Nostradamus or Edgar Cayce, the later also a member of the "Disciples of Christ." Many people question strict Orthodox law, an attitude that gave rise to Protestants, Lutherans, and other groups that objected to the Roman Catholic Church and its many traditions and rites. In the early 20th century, the Stone-Campbell movement was associated with the Second Great Awakening, or re-examination of Christian doctrine. All these different views can be confusing; trusting in the Bible and working out your own reconciliation to God is suggested to be up to the individual. For centuries, the Roman Catholic Church was considered the source and power of Christian tradition, but many protested its excesses and abuses…hence the term "Protestant." Likewise, the Stone-Campbell Movement, which led to many "New Age Movements." The Lutheran Church came about from Martin Luther and numerous theological documents. Cayce's "Disciples of Christ" merged with the "Churches of Christ," and formed the "Springfield Presbytery," which disassociated themselves with the "Augsburg Confession," one of the founding documents in the Lutheran movement in 1530, and was the fourth document of their "Book of Concord."

Understanding the back and forth movements throughout history is confusing, and shouldn't detract from the basic outline for reconciliation described in the Gospels.
I believe we need Jesus for redemption, and true repentance involves honest contrition for sins, and for the faith laid down in the Gospels, thanks to the words of Christ. The Catholic confession, when you vocally list your sins to a confessor, and many of their other Orthodox teachings needlessly complicate what should be a simple act or repentance, redemption, and faith.

Our "New Age Movement" is a complete cornucopia of meta-physical traditions, Holistic health, parapsychology, consciousness research, self help and motivational psychology, and quantum physics. Such a hodge-podge of beliefs, all designed to create, "A Spirituality without borders or confining dogmas," and empower the Mind, Body, and Spirit, is too far out there to even consider. Embracing mainstream science, the Age of Aquarius, and a worldview that attempts to blend what's new with what's old appeals to some, yet could never replace the simple words of Christ and the "Good News" outlined in the Gospels. No wonder people get confused.
Time to take a little break from the history of righteousness…Whew.

The following is a story from 2003: developed, but not finished, edited but not polished; it is what it is…possibly a novella, or a short story…it touches on new age wisdom and belief, but also shows goodness triumphs. Science is able to photograph our auras…takes a special camera, but the results are interesting…it hasn't been used much, but I'm sure some auras are vastly different, and are recognizable signs of what you're like. Perhaps that's why UFO's watch certain people.

Weird Sky and Indigo Eyes

Kendra ran up the front walkway, spun around as she opened and closed the front door in her usual fashion, continuing up the carpeted stairway until she reached her favorite sanctuary, her newly decorated bedroom. She tossed her book bag on a chair, removed her running shoes, and jumped on her bed. Still angry over the latest teasing from Vicky Gambrell, a cheerleader & air-headed bully, she tucked her legs under her chin and thought about the discussion. Vicky picked on her relentlessly, getting entire groups of kids to laugh at her. Sometimes it was too much, and she just wanted to stay home, anything to escape the abuse. Kendra knew she should not get upset over trivial emotions; she was much more mature—self-educated and beyond the awareness level of most people.

Her uncle Jim taught her as a child, telling her she was an indigo kid; quite special. Jimmy was a PhD that taught weird theories on how the world's many intersecting force fields worked in harmony. He said he was writing a paper on the universal theory of field unification: electricity, magnetism, gravity, cosmic rays, bio-luminescent transference, electro-magnetism, microwaves...he said they are all around us, and we could see them if we had the right vision. He said all living matter emits an electrical field called an aura, and that humans had really special auras, as they were produced by our souls. Mastering and extending this field could manipulate other fields, something that required an innate ability that he said Kendra possessed. It sounded neat, but it was too complicated for Kendra to understand. She asked him about thought, and he agreed it was different, and that it might be like a projected beam. Kendra told him thought was faster than light, as all she had to do was think something, and her thought waves could be wherever she projected them. Her uncle smiled, and called her a really special kid.

Nevertheless, her 14-year old hormones and teenage emotions could interfere and interrupt her infinite understanding and channeling ability of the vast, universal, and minuscule level governing vibrations—to focus and allow that energy to flow through her and be directed to continue the galactic relationship we know as life. Just how this understanding came to her had never been a mystery; it was just a perception that grew in intensity until she became physically conscious of a new level of enlightenment and alertness. She thought of it as if someone were in a dark room, almost pitch black, and a small light source provided enough ambient light to see your way around the room, if only in quick flashes. Similarly, in a crowded dark room, a flash of light would engrain the content's location on her brain, allowing her to walk around without stumbling.

She closed her intensely blue eyes and began to meditate. A process she discovered allowed her to accept her condition and not be overwhelmed by the occasional petty incidents that made up her everyday life at school. Meditation allowed her to focus her thoughts, and once her mind was clear, she directed her thoughts towards the sky, right where the planet Jupiter continued its slow transit of our sun. The last time she'd tried this, she received a response; only after weeks of thought did it begin to make sense. Now that she knew what it said, she wanted to convey that awareness to whatever sent her the thought, and to find out more. Her mind's eye pictured her thoughts as a beam of life, searching the black void for something that had substance…something that had intelligence. It didn't take long. Another thought beam intersected her's, and a two way exchange was possible. She immediately communicated what she had learned, a conclusion that pleased whatever she was linked to.

Now more thoughts were sent her way…slowly at first, but speeding up, eventually becoming too fast for her to understand. After a long burst of unintelligible data burned into her brain, as softer thought, slow and easy to understand, told her the information would unfold, and she would soon be able to exist with them, in bodily form. Kendra understood they were sending for her; the knowledge they shared was coalescing in her mind, flowing through the tendril of spidery neurons that traded thoughts and memory in electrical impulses. She was quickly sorting out the information they shared…it could be their language, their history, or advanced science. Whatever it was, she knew she would comprehend the message. The craft would be there when she was ready, and that would let her ponder the message, letting the data unfold and flow through her mind. Just before the communication was terminated, she received an image of life; beautiful, abstract—a representation, it was bathed in an indigo light, and it was the most remarkable thing she had ever seen.

The image still fresh in her mind, she decided to paint what she imagined. Art had always been a passion for her, and she was talented and skillful with a brush. Getting everything ready, she began to paint. Time seemed to stop while she painted, as her mother finally burst into her room, telling her dinner was ready. When she saw the picture, she gasped. It frightened her. Running from the room, she yelled back, "What the hell is that thing, it looks like it's alive."

Kendra looked at the painting with love, but could see nothing but life; existence taken down to its fundamental level. A universal existence that had no limitations, no restrictions—this was life that both above and below, the simplest and the most complex. It was lovely.

She joined her family at dinner, and woodenly joined the conversation. She wanted to leave and finish her painting. She passed on desert and ran up to her room. The painting has somehow moved closer to the window, as if it was looking out the window. Kendra didn't mind, as it gave her better light to work with. She picked up her brushes and continued to paint; each extracting detail, every minute line and shape. The closer she came to the final image in her mind, the more the picture seemed to develop a life of its own. Features changed, then return to normal. What could be eyes moved. Kendra was finished. She placed the picture in front of the window; if it wanted a better view, why not?

Tired from her painting marathon, she brushed her teeth for bed. When she looked in the mirror, she saw her eyes were now deep pools of indigo. Cool, she thought, jumping between the sheets. She was out like a light.

Morning came, and she woke up with her brain buzzing—the entire message had played out during the night, and Kendra knew exactly what it was. An inter-galactic message…a download that all who were able to access through unused areas of their minds could understand…this message was for them. She looked at her picture. Out of the easel, it was flat against the window; she now understood the painting was really a hyper-dimensional object that was a communications medium, channel thought back and forth. She opened the window, and the painting flew up into the sky and formed holographic rectangle in the clear blue sky. The intense indigo was muted by the sunshine, and the sheer size and dimension of the object was enormous, but hard to see. She understood only people with indigo eyes, soulful eyes, could see the object, and only they could see what was in the object.

Between the clouds and the horizon, the picture existed…it showed a picture of a strange planet, with enormous, geometrically designed buildings. Some levitated in the strange blue sky with no means of support. Kendra realized it was a gateway, a passage-way to that world. The sheer beauty and perfection of the pyramids, octagons, tall rectangles mixed with other shapes, gardens and waterfalls everywhere, a large geodesic domes emitted a soft blue light, warm and light, especially against the indigo background. Square pyramids stretched into the light blue sky, triangular prisms split the light into specific beams, the different colors falling upon another shape that glowed and pulsed with that color. Kendra instinctively knew they were power supplies, a clean power of colossal proportions. She could make out smaller shapes, flying around, entering solid buildings with no openings, as if the surface was fluid. Other fantastic shapes appeared everywhere, all functional, and emitting waves of emotions…joy, happiness, and love; the many trees and plants seemed welcoming and alive, the crystal clear water infused with energy as it followed impossible looking courses.

The picture had now taken over the horizon. She went downstairs to get another look. Her mother stood in the doorway, her mouth agape, and said, "Wasn't that what you painted last night…how…Kendra, what's going on. Kendra gave her mom a loving hug, and quietly said, "It's all right mom…it's what we'll all wanted, love, true love, has finally visited Earth. We should feel honored." As her mother gazed at the startling scene, Kendra walked over to the top of a hill to get a better look. Sunshine now shone though, but with a light blue hue. People were all out of their houses, looking at the sky. Some were holding out their arms, as if beckoning something to come through.

Beams of intensified sunlight flicked and flitted about; falling on some groups of people, the people instantly disappeared, as if the light had transported them somewhere. Excitement and joy was in the air. Looking around, Kendra realized this horizon spread across the whole world, and countless groups of people were being taken by the beams.

The peaceful scene was soon shattered by the screaming engines of F-18's, all heading off in the distance, as if trying to capture or enter the holographic scene. It wasn't a doorway or portal, thought Kendra, it was a projection formed by the hearts and minds of all the people she must have touched and awakened in their sleep. The welcoming crowds with their arms held out were touched by the sun beams and quickly vanished. She imagined this was happening all over the world. More people were emerging now—some were half dressed, and some seemed surprised and angry. They pointed at the scene, and it looked like they were yelling at each other. Watching the events unfold, Kendra realized none of these people were touched by the sun beams, they just stood, angrily shaking their fists at something they couldn't understand, and therefore something that frightened them. It seemed like the entire city was out watching the weird sunrise; standing on tops of buildings, houses, and large hills, Kendra could see the large groups, some of them pushing and shoving. Each time beams of blue lights played over them, some were taken, others not. It was like a Biblical event; when Jesus said: "Two will be in the field; one will be taken, the other left...three will walk together, but one will remain." The beams were taking the humble, the good; rich men stood on top of their business, shouting at the light, for the word had been spread that the light was taking the elect to a higher plane of existence.

Explosions could be heard; the fighters had launched Sidewinders at the strange horizon, according to orders from the puzzled and irate commanders. They couldn't imagine what this was, and could only perceive something as a threat, or something they could control. Thankfully, the spent missiles fell harmlessly to Earth, detonating in a deserted forest or open plain. In the vast spectrum of lights, people's individual auras could now be seen; some light blue, some white and bright, some light green...others were fiery red with anger, others black. Whoever wasn't a light hue was a shadowy shade of the darker colors...inky blue, deep brown, dark green; here and there, bright red colors fluoresced and pulsed...their auras seemed to stretch out from them, just as the white auras shone brightly, and seemed to shoot out of their arms when outstretched.

Kendra understood what was happening; the image she created was now a portal...a doorway to that beautifully blue planet—light and bright blue was interspersed with wonderful shades of green, the trees, the grass, and the plants, as clear streams and waterfalls caught the soothing light and reflected it across the image that now stretched in the distance, no matter which was you looked. People's auras allowed them to be selected; the dark auras were left behind, as the lighter auras floated up on the light beams, fading into the light in mid air. She could feel the emotions of the people surrounding her. Love and joy were mixed with envy, hate, and jealousy...the bad emotions weighed these people down, and they were unable to float into the light, as they were too grounded to the Earth. The rich, the powerful, the hateful, and the violent...all the negative emotions were being weeded out, and only the carefree people full of love and hope were riding the light to a higher existence. Kendra imagined they had indigo eyes, like hers. Fights had broken out on some of the rooftops; the dark people were trying to jump on the light people, but fell back into the dark crowd who were all becoming agitated, worrying that this might be the prophesied second coming of Christ, or some other Biblical event they knew they couldn't pass.

They knew the people caught up in the light...the meek, the humble, the kind, the milquetoasts—the sort bullies picked on—busboys they just degraded and ridiculed only last night. Now they were wispy, whiffles of air, rising above, evaporating into the light. The bullies didn't know where they were going, they were jealous they couldn't come along. People in general were fed up with this world; leaving it in an ethereal, magical manner seemed preferable to returning to the endless fight for the almighty dollar...stepping on people to make money. Somehow, these "undeserving" sissies were somehow getting selected for something incredible, and the dark people didn't like that. Chaos was breaking out, as more groups no longer had the people of light—they were all a dismal black...an inky shadow that showed their true nature, thanks to the distinctive aura that surrounds everyone.

As Kendra watched the horizon, she saw a pinpoint of light coming towards her, gradually growing bigger as she watched. Soon a bright disk hovered over her, totally silent. A bright beam came down, and she felt her body grow lighter, the tug of gravity no longer pulled her down...an encumbrance that seemed to fall away, freeing a fluffy version of her that was pure spirit. She turned and looked at her old house. Her dark brown mother stood on the steps, her expression hard to make out. Kendra thought she was crying, and reached out to her. A ball of light left her hand and hit her mother. Like a high-beam flashlight, her mother was caught in a ball of light. The radiance flowed over her, gradually removing the dark. When she last saw her, her mother was now full of light. Kendra was bursting with happiness and love. Her mother shed her darkness and would join her, wherever that might be. She was now on the ship, a vessel with transparent walls, and other beings that seemed like transparent shadows of light. A being with a pastel-blue aura approached her, and she immediately knew this was who she had been in communication with. Words were unnecessary as she conveyed her gratitude and thanks. The being surprised her and knelt before her, indicated he was honored by her presence. Kendra reached out and wrapped her arms around him; it was an intense feeling of incredible love and happiness. Their auras joined together and gave reflected off the shiny, transparent windows in a frame of silver. She asked about where all the light people were going, and he confirmed her conclusion they were joining the planet that appeared across the world's horizons. He also confirmed they would all understand each other by thought.

In a brief, but intense flash of insight, she realized they'd all reached a higher plane of existence, one where the power of their minds could manipulate matter on the subatomic scale, and harness incredible sources of power through diverting free energy that floated around the universe in waves of cosmic particles, light beams, and the immense gravitational power of a black hole. She also understood the eternal function of the universe: black holes created other universes that appeared at the end of the vortexes engulfing matter and energy in a never-ending sequence of creation and recycling. The moment passed, but for an instant, she could feel the incredible wisdom and power of the strongest force in the galaxy, the omnipotent force of goodness, God. Everything was in balance, and she understood why the heavy darkness existed, but realized she had now evolved into a much higher form of being, and that the dark forces would never again come near her, or they would be destroyed by the much stronger force of goodness and light, like sunshine pushing aside the cloak of night.

The ship seemed to sit upon the world, reflecting all sides at once. Only the shadowy people remained, and she realized that over time, they too would come into contact with light, allowing themselves to rise above their condition and transcend the world.

Day 2:

Mull and absorb, meditate and matriculate

Ignorance and immoderation, two backward principles that created a dark age in scientific advancement. How far ahead would we be without this shameful period?

The dark ages can be seen as the result of an over exuberant Church, a structure some say was faulty after the 325 Nicene Creed was promulgated, and the Church started competing with Kings to get small donations from a impoverished lower class—Kings demanded taxes, but the Church offered "Indulgences or religious artifacts," saying you could "buy" your way into Heaven. Anyone that didn't tow the official Orthodox line met the Inquisition, and they usually burned you or some other nasty death penalty. Sadly, many people were killed so the Church could take over their lands, get the water rights they needed, and other selfish motives. They even burned four Franciscan friars because they refused to own property...that really shocked me. Torching four righteous monks, just because they wanted to follow in Jesus' footsteps. The drastic and long term effects were on scientific advancement. The Church was quick to label anything they didn't understand or like, "The work of the devil." With such an attitude, many scientists studied secretly, and those secrets were often lost. I sometimes wonder how further society would be in technical progress if scientists were allowed a free hand to pursue empirical experiments, understand chemistry, basic physics, and astrology.

This story is based on something that might have happened, and if it did, it was a crime against humanity...keeping society in the dark, just so they could control a gullible population and keep those "tithes" and donations filling up the collection plates.

Historical loss: We repeat what is not remembered or recorded for future reference.

Story outline/start:

1248 - Glasgow Scotland

Their existence and beauty puzzled him. Strewn across a pitch black sky, dazzling, twinkling dots of light shone with mystery and appeal, promising many answers, but posed even more questions. On such a cloudless night, the sky looked like a backdrop of silk with thousands of pinpricks in it, letting starlight shine through. It's beauty was undeniable, a creation worthy of God; based on his rudimentary calculations, these objects were extremely far away, far beyond Aristotle's model of a perfect balance of spheres, with the Earth at the center. He was sure that was wrong; one night, he witnessed a round shadow shift across the moon—only the shadow of the Earth could produce such an effect, and that would mean the sun was central to the planets, the true axis around which everything revolved. But, even mentioning such an idea would get him burned at the stake by the narrow-mined inquisition, they treated real science like magic, and refused to see what their eyes told them was true. Blind faith, or so they said, but no where in the Bible did it say the Earth was the center of the universe, it was merely another faulty man-made law or interpretation, something that held humanity in ignorance. Using empirical science, he had developed a theory that he couldn't share.

Alfred Macdonald could remember thousands of fanciful descriptions of the night sky, but he believed he discovered the real nature of the holy heavens: many of the larger lights were planets, revolving around the sun, several having moons just like ours. He would have been charged with blasphemy if he revealed the church was wrong, that Ptolemy's geocentric model was wrong. Trying to get a closer look at the sky, he extrapolated on the recent invention of the magnifying glass. Reasoning that if convex glass magnified objects, then the opposite shape, concave glass, or a combination of them, would have the opposite effect. Using this principle, he tried different glass shapes and positions until he had a working "distance tube," an instrument that allowed him to see across great distances. He noticed the battered surface of the moon, and realized it revolved around the Earth: from there, it was easy to see other objects that followed similar paths, all revolving around a central point, the sun. A heliocentric model made more scientific sense than the imaginary rules by which the church lived.

Prior to his discovery, the star's existence and beauty always puzzled him. Across a pitch black sky, dazzling, twinkling dots of light shone with mystery and appeal, promising incredible answers, but creating more fantastic questions. Whatever their function, their beauty was undeniable. On cloudless nights, the sky was a backdrop of silk speckled with powerful glittering dots of mystery and intrigue.

Alfred Macdonald could remember many fanciful descriptions of the night sky, but he knew the real nature of the holy heavens: many of the lights were suns, likely with planets revolving around them, just like ours. It was knowledge that could get him killed. It was considered blasphemy to even suggest the church was wrong, and that Ptolemy's geocentric model was wrong.

When he turned his invention on the heavens, he was stunned. He saw mountains and deep craters on the moon, observed sun spots and realized that some of the other planets had moons orbiting them. He knew his work would change the entire world view, and discredit the 'perfect spheres' theory of Ptolemy and Aristotle; most of all, he was terrified of the Church Council and self-proclaimed soldiers of the faith.

Fueling his paranoia was the local Bishop Eugene Cornelius, a stern and fervent servant of the Church, who zealously persecuted heretics and passionately pursued any deviation from the teachings of the church. He was suspicious of any scholar who did not concentrate on biblical matters and had a closed mind when it came to science. The Bishop believed science and alchemy were too closely connected; therefore, anyone who deviated from biblical studies was the tool of Satan. One of his oft repeated phrases was 'God has revealed everything we need to know, and faith provides us with all other questions'. To a questioning scientist, this view of the world was a death knell to further research and empirical analysis.

Night after night, from the top story window of his apartment in Glasgow, he searched the heavens, making new discoveries and entering them in his journal. He saw the four phases of Venus, something impossible under the geocentric model, and discovered four satellites orbiting the largest planet after Mars. Alfred continued to make more discoveries, all dutifully recorded in his masterpiece. As a natural philosopher, he knew the recognition and respect his work would bring him, but again feared the wrath of the church. This was not the time for change. The Catholic Church was too powerful and influential.

Making the greatest scientific discovery since the great Greek Philosophers, he had no idea that his temerity to act would result in another 250 years of darkness. When his work was repeated by great men like Kepler, Copernicus, and Galileo, there was still controversy and resistance from the Church. Unknown to him, his discoveries would be duplicated by another man, with a device called a telescope, but hundreds of years after his discoveries. Galileo, a man brave enough to stand behind his evidence, would have benefited from his knowledge, but Alfred was too afraid to share what he learned. He kept his notes secret, and only trusted one man with what he had found: William

McTavish, a close associate of Sir Roger Bacon, the empirical scientist of Oxford, but also a member of the Franciscan Order. Alfred heard of Bacon's commitment to science, and hoped this man would benefit from his work, or even share with other scholars the truth of his observations. If others could look through his distance tube, they would believe. He hoped McTavish would travel to Oxford and show the teacher what he had created, and what it opened…a universe of endless beauty and wonder. Alfred knew this was the right thing to do, but his cowardly nature always surfaced, convincing him someone else would be better suited for this endeavor. As he grew older, he knew he wouldn't make the journey to Oxford himself, and hoped the very nature of his discoveries would push men to act. Science would be catapulted into the forefront, as it always should have been; having the church decide what should be studied was a colossal mistake, an error that trapped them all in their narrow view outside the Bible.

In the last years of his life, Alfred finally had the courage to send his complete book, called The Cosmologia, along with his distance tube to William McTavish, Earl of Warwick, one of Scotland's foremost scholars and men of science. McTavish was close the Sir Roger Bacon, and perhaps together they could propel mankind forward…taking a step beyond the dark and confined world that the church insisted was the only orthodox view of existence—anything else was considered to be of the devil. He trusted this man would know what to do with the detailed information and drawings. Alfred died shortly after, never knowing the results of his life's work.

A crime to progress, when McTavish received the comprehensive book, he merely glanced through it, and concluded it was a hoax. He didn't examine or even bother looking through the tube that accompanied the book. Having a large library, he placed the book and tube among his curiosities and carried on with life. The House of McTavish prevailed throughout the following centuries, the original castle always part of the McTavish's holdings. The library remained intact through generations of his descendants, some added to it, but no one examined it carefully enough to discover the extent of the Earl's original collection.

2008 – Present Day

Warren McTavish arrived home from the University of Glasgow, where he was a Professor of Ancient Studies. Since arriving from America, he'd been browsing through the old library that had been in his family's hands before Robert the Bruce ruled Scotland. While looking for something on the third tier, when he found a cabinet he had not noticed before. Intrigued, he opened it and discovered an old, handwritten text that was hundreds of years old.

Quickly scanning the book, he could not believe what he was reading. He was holding proof that someone had predated Galileo's discoveries by roughly 250 years. He rummaged through the cabinet, full of old letters and documents, and found a crude telescope, no doubt of the same era. "What a crime," he thought, "religious doctrine forced ignorance on the world throughout the Dark Ages." Suppressing this knowledge caused man to live in ignorance for centuries. Realizing the book's significance, he wondered what circumstances could have buried such an important scientific breakthrough for so long. Realizing this would re-write history, he began to phone and fax select people.

While reading the letters from Alfred Macdonald, he noted a letter to his great ancestor. What he read made him sit down in shock and despair. The scientist that discovered this advance in astronomy had written to his forefather, and trusted him with his discoveries. Scrawled across the bottom of the letter was the death knell of all scientific progress, and his family seemed as guilty as most. It said: Sacrilege…the sun revolves about the Earth. Any fool can see that from watching the sky. This is the work of the devil. In a larger hand, the word SCANDEL was scrawled across the bottom. Many doors were closed in those days, Warren lamented, but the real crime was closing their minds.

Fictional, but I wonder if similar scenarios occurred…scientists are curious, and can't help but explore the unknown…it was just sad that when true discoveries were made, religious minds, not scientific minds, were in charge of new discoveries, and their religious blindness kept us in ignorance…a true crime that went on for as long as the Church could get away with it—a crime of the mind.

DAY 3:

A day of discombobulating despair: I foresee a lot of these

I lived with sorrow, I lived with hate. Forlorn and dejected, my life was lonely and an abysmal failure; I was spiritually dead, crying out for love. It's time like these that make us look for spiritual help—when all you have is God, you'll find that God was all you really need. Faith and acceptance sometimes requires us to scrape around the bottom of the barrel; when thusly humbled, you look up, see redemption, and realize redemption is attainable, but it needs a truly humble repentance. When you've lost everything, a fool rages against God for taking from him, but a wise man understands that God only removed the things that blocked your view of Him, and rightfully removed them so you could really begin to live, with God as your mentor and companion.

I had no outlet for my sorrow, no reason to live. These words are often said with many tears, when your life has been stripped of the unnecessary baggage that was weighing you down. A prudent reaction is to examine your soul, find out what's wrong, and wash it in the cleansing blood of Jesus. Your sorrow will turn to joy, and you'll be blessed with a new reason to live. This is only possible when you open yourself and take a ruthless inventory; usually, people find they're lacking in spiritual enlightenment, realizing that's the only way to rise above the sorrow, renew yourself with new beliefs, and begin a quest for spirituality that will be rewarding and revitalizing. Rid yourself of old values, and embrace something that is akin to moral magic. If you are blessed with experiencing the Holy Spirit, it's like something strange and wonderful has happened to you; you're shrouded in a cloak of pure love, and an indescribable joy leaps within your heart. True bliss beyond human understanding, yet merely a taste of what awaits the righteous.

You realize human values are meaningless next to Godly wisdom and power. When you think you are wise, your foolishness blinds you to the truth;

grasping for solutions, you're wise if you start on God's path, for that leads to life and true happiness. Wallow in pity and despair, or humble yourself and receive honor; it will be the best decision you've ever made. A small dose of heavenly joy can remake you and alter your outlook forever.

Thankfully, I picked up a Bible, remembering the only time I truly felt love - when I was seven years old – when I was called to give my life to Christ. Over the years, I forgot Jesus, but Jesus Christ never forgot or gave up on me…lucky for me, but that's what true love means…the power of love is a lot more powerful than we give it credit for…we often just think about personal love, like loving a wife or girlfriend. Loving the world, not the "things" in the world, is like a state of awareness. Loving all penetrates through layers of hate, regrettable emotions like envy or jealousy, and gives you a universal feeling of love that transcends the Earth, becomes spiritual, and is universal. Somehow the force or "essence" of love exists throughout the universe…perhaps that is what makes a flower so beautiful, or a poisonous vine ugly—scientist talk about figuring out a "unified force theory," but how many forces are there? We've only begun to capture a person's "aura" in photographs, so who knows what kind of forces exists…and how they all interact with each other to keep a coherent balance throughout the universe. Pictures at eleven.

Reading the Bible, about His life and His sacrifice for me - a hopeless scoundrel, I fell to my knees in prayer, crying out for forgiveness and redemption. Tears of joy replaced a broken heart, and I totally surrendered my life to Jesus Christ, my Lord and Saviour. Words are not enough to express my gratitude to Jesus. Try it, you might like it.

The Word of God restored my faith in Jesus Christ, giving me a guide to live. All questions, all my doubts, are now answered. Jesus allowed me to live again in Him, and to love.

I am still learning, but I now feel God's love and presence in my life. So much so, I believe I understand what "Transcendence" and other states of advanced awareness might be. Our short life will soon be past, and only what Christ has done on the cross will last.

I have a picture of a UFO that was hovering in a tree, about 200 feet in the air, and I was the only person awake and around for miles, as I was camping in Manning Park, B.C., a very remote spot in the wilderness. The colors of metal, the lights and the shape of the craft are quite detailed, but I only saw it in the picture I took of the moon, a few weeks after the fact. I wonder if these things can make themselves invisible to our senses, but are unable to escape the mechanics of a camera.

There were also weird "orbs," or spots in the shot, which might have been inter-dimensional doorways, as one had stuff visible inside, geometrical shapes that looked like buildings. I do not pretend to understand what I captured on film…it's something very strange, and I would show it to anyone who thinks they might know what it is. One fact is very weird. It was after 1:00 in the morning, and I was enjoying solitude by the Similkameen River, so why was it there, I was the only human being around. Were they watching me? That is as disturbing a question as the picture itself. Why me? Because I've attained a special "aura," an aura that shows when a person has attained goodness inside, and projects a specially colored aura? I've seen pictures they've taken of the human aura, so I know it is a scientific fact, but I don't know if they are different and can depend on what level of personal awareness or enlightenment a person has reached.

Peter 1 5:7 "cast your anxieties on him, because he cares for you."

Day 4:

Another day in the year: must have been a writing day, or could have swallowed my big mouth…

Everyone has their own standards; their own likes, dislikes, and pet peeves. Whether it's an attitude, an overused clique, innate behavior, prejudice, or a nasty habit, everyone has personal traits that can be objectionable to others, but are justified and seen as perfectly natural by that person. Despite his own imperfections, he ignores them, assumes he is above reproach, and free of bad habits; he then uses this selfish arrogance to judge the bad behavior of others, fully convinced he is blessed with impunity. Hypocrisy is thus displayed with flippant disregard for personal integrity. As Christ so wisely admonished us, "let he who is without sin cast the first stone", a phrase so aptly put that it shows us we are all incapable of making moral judgments against each other. We are able to rebuke offenders who have broken our civil laws, an entirely different set of standards that should be left to lawyers and elected judges to deal with.

(See: The Document Legacy: mysteries of history…novel)

Personal judgments are made every day, everywhere, and against every person. We see what we want to see and it offends us; we therefore make a judgment and unfairly treat another human being for sins, lifestyle, or behaviors that we do not fully understand. Who then is the worst offender here? To judge a man by the clothes he wears, or to ask him if he needs help because he might look like he was in need of money. The replies might startle you. A person could be extremely rich, wearing tattered clothes because he is doing a chore that would get his usual clothes dirty. Someone could dress that way because of preference, or some later-day Grunge/Hippie fashion statement. And yes, you might find someone suffering from poverty and who would appreciate it if you would help him out. We only live with each other as strangers; even strangers still have the same needs, wants, and necessities. Helping each other is the only way we have to show each that we care about our fellow man. Too many people just think about themselves; they ignore what they don't like, avoid what they are ashamed of and close their eyes to situations that we find disturbing. That is the mark of hubris; that a mark of selfishness, a stigma that people are quick to find in others, but completely ignore personally. Ultimately, being judged by the same standards someone uses to judge another person is the fairest way of finding merit. When someone abhors something they see everywhere around them, verbally lashes out against it whenever they have the chance, it's shameful that person can't recognize the same behavior in himself. Jesus said, "Judge not lest ye be judged… the yardstick you measure others will be used to measure you…or you'll be found guilty by the same argument you use to judge everyone else."

"How can you tell a brother he has a sliver in his eye when you have a log in your own?"

"First remove the log in your own eye so you can properly see to tell your brother he has a sliver in his eye." Don't blame someone unless you're blameless…and Christ showed us that we are all full of blame by living with sin. Only through the grace of God will we ever be forgiven, and we must have faith in Jesus to intercede on our behalf before God. Without Jesus, there is no way to the Father. For it was said, "I am the way, the truth and the light, no one shall see the Father except through me."

Beware, you have been told. Don't put off until tomorrow what you should be living today. Joy enters your life, and you must learn how to become Godly. It is not easy; it is not an instant solution. We were put on this Earth to suffer and learn. Living the high life here makes you forget about the riches to come. Anyone fixated on this world will not see the next world. Read your Bible and you will find out.

Beyond return—over the edge. One toe over the line. Treading a circular path. Aimless and pointless. Everything yet nothing can convey the pain and despair of watching your life pass you by; you are just an observer, a passenger that has little control over external circumstances. Generations of people all suffered the same afflictions, but most of them are unaware of the waste until their last few seconds, when their life is supposedly replayed for them in a flash. What if you wake up prior to the end of a wasted life, realize you want to change everything about yourself, and try to make a difference in this world. You suddenly discover you must do anything that would help other people, give your time and attention to people that need it, or give what you have to others instead of selfishly wasting it on yourself. People call such instant discoveries epiphanies.

Day 5:

Ideas become dreams, and tend to stick around

The life we live is up to us; unless we understand the sacrifice Jesus Christ made for us, we stand alone in our sin. Our sins are ours; our forgiveness is only available through faith in our Lord and Savior, Jesus Christ. What humanity doesn't understand is that God loved his creation, the human race, and carefully made his presence known to us after our forefathers introduced us to sin and evil. They were so infused with corruption that God had lost hope in Man. After creation, he gave Adam and Eve a very easy way to live. He only asked one thing; not to eat of the tree of goodness and evil, for it will surely destroy our innocence and make us unmanageable and difficult to love. Our God is a God of love; asking for our love should have been a joy to give him, but our curiosity, along with the manipulation of an evil serpent, introduced us to pain, corruption, and evil. He made several covenants with humans, but they ignored God and forced him to take action. Yet, after every time he punished us, he gave us a second chance and outlined what we needed to do to worship him and endear ourselves to his loving embrace. He almost wiped out humanity because of our evil ways, but found favor with a few individuals who managed to appeal to his mercy and thus allowed us another chance and new covenant with our God.

Finally, in His ultimate wisdom and incredible planning, he gave us His only begotten Son, Jesus Christ, that would act as an intermediary between sin and forgiveness. Seeking grace and righteousness through our Lord and Savior, Jesus Christ, we join him on the cross so that our sinful bodies and spirits might die with him and be reborn and purified through his blood. He atoned for our sins through his physical death, but his mighty power and grace conquered death and lives on in a pure form. If we accept his teachings and ask him into our hearts, we can begin to live a sinless life by following his examples and teachings. We are always tainted by sin, but through his mercy and grace, we are forgiven, provided we live our live with faith in his message and blessing. We can begin to live a Godly life through trust and faith in our Lord. We can be saved if we live in the light and avoid the darkness of sin.

A moment of clarity that blazes into your awareness like a blast of sunshine on a cloudy day. Your mental facilities, or what is left of them, may understand the importance of performing helpful acts of kindness. They would allow you to feel the experience of giving, a wonderful alternative to your previous history of taking, hoarding and stockpiling things of value. You want to give to others; unfortunately, society has many rules that protect the needy, making it hard for someone else to actually give them something. Organizations that hold charity up as its reason d'etre almost monopolize the entire process of giving. In some cases, these charities screen whoever tries to give them free labor, complicating the entire process.

There are some people that create an entire system so that they can be seen giving their personal time and effort to help the needy. They do it for themselves, not for the people that need help. The poor eventually get the occasional donation, but only after standing in line and going through a complex process that distances them from the help they so desperately need. They are more concerned about being seen as helpful human beings rather than actually being helpful; everything is about them is false, even their personal self image. They overlook their bad habits and convince themselves they are good people. Their life is so full of falsehoods and lies they can't tell the difference between reality and make-believe.

When someone's eyes are open to the truth, it can drastically alter someone's perception, attitude, and behavior. If God gives the truth to that person, they are often called for a specific purpose. God can also call someone but be somewhat cryptic about what he wants that person to do. We sometimes don't need to perform a specific function, such as lead a multitude of people to the path of enlightenment; our behavior alone can be influential to a confused person, or we might tell someone what he needs to hear to finally get back on the road to God and Christ, the source of all goodness and love. As humans, finding Jesus/love is the ultimate purpose in our life. Sharing that with others is important; the more believers we bring to God the better the world will be. There is a apocalyptic series of events that Jesus told us to expect in the final days, but loving our Lord and following his example is the best way to lead our lives, and not worry about end days, or massive final battles. Be aware of which side you're on, and that should be enough. I you want to join Satan…kiss your ass goodbye, as I'll be on the good side, and I have a feeling I'd make a formidable Heavenly warrior…then again, who knows what will happen, only God knows, as they say. Perhaps humanity did something to change his mind from the days of John of Patmos and the divine Revelation of John; brutal war, major cataclysms, the ocean turning blood red, comets hitting the Earth, the Four Horses of the Apocalypse…it all sounds frightening, and it might all come to pass, just as predicted. However, with all the attention on Life Ending Events after 2012, everyone knows what is supposed to happen…so who's going to believe an anti-Christ, and get 666 stamped on their head? That would be like asking to go to hell, knowing you'll roast for eternity, and do all this with the full knowledge of its Holy repercussions?

It seems to be an unfair system when someone that was at one point on the fringe of society suddenly wants to change and join society, adopt their rules and obey the law of the land, but encounter prejudice and denial. Society still views them as a pariah, someone that is unworthy of joining the good guys. The stigma of past sin is never forgotten, even though society makes a big deal about extending forgiveness. There is a point to some of this, as anyone who was once corrupted by the lure of evil has a difficulty ignoring what once destroyed him or her. The possibility of relapse, the call of the wild, is strong; it takes a lot of faith and personal commitment to overcome what once captured your soul.

But what of the person who really wants to change their life? Are they never to be trusted? It can be better to extend a helping hand with a welcoming heart. Let the person who wants to change see the benefits of living a good life. If that feeling of compassion and love is withheld, the person who wants to change encounters barriers that only add to their problems. Extending trust might backfire, but it also might be the support that person needs to fully forget his past and learn to live a life of goodness and care.

If someone were addicted to something that was powerful and difficult to overcome, that person could use all the help they can get. That help might make the difference between securing himself on the road to recovery, or sensing the hopelessness, he believes lies between him and sincerity. When someone is climbing out of a deep pit, a hole with slippery sides, and a bottom that tries to pull him back in, offering a helping hand can be the only way he is able to free himself from the pit. That is what should be done if you really want someone to exit the pit and restore himself to normal, level ground that everyone shares.

Day 6:

More on yesterday…a mental roller-coaster?

People can sometimes look at the poor and poverty-stricken and think they are better than they are; part of this can be considered true, as the people who are living a good life are right. They follow the law and are good citizens. Having a good family, a good job or a special position in life rewards them. The person suffering from addiction, homelessness or abject poverty is paying a price. They live in misery. They don't have good friends or a loving family. They are often victims of people who stepped on them so they could benefit from them. People often take more than they need when they should have shared what they had. Greedy relatives who won't share an inheritance or give someone a place to live sometimes put the down and out in the situation they find themselves.

It would be an act of kindness to treat the unfortunate with open arms and love when they really deserve it. Many of these people are greedy, selfish, and only concerned with living the way they want to live. They are were they belong. They don't want help. They want to feed their addiction and continue to live outside the law, and live the way they think they should. So, be it. However, when someone expresses the desire to change and takes steps to do something with their lives, they should be forgiven and given the chance to join life on an equal footing with those who control society. Everyone is equal in spiritual matters. Deeds always speak louder than words, but good intentions are untried possibilities. Someone should give people a chance. A chance to learn the correct way to live, along with the chance to fail, if that is what will happen.

What seems like the main idea here, according to the hindsight editors, is that we should give people a chance to join the world without prejudice. Not everyone can clean him or herself up and go on to be a person of import; they can, however, become who they are meant to be, if they are given the support and opportunity to fit in. Sadly, the good opportunities are guarded jealously and only given to people who know someone that can get them special preference. People often cry "foul," or "unfair" when someone is given a chance that normally would go to someone that is seen as deserving of it, or has earned the chance. No one wants to take a chance on someone that has a history of failure, but in reality, that chance may very well give someone the ability to realize they need to straighten up and perform the job they were entrusted with, thus allowing them to become a new person, full of appreciation, and devoted gratitude. That appreciation fills them with gusto, and they work harder than the average bear. They realize they are indebted to society, and might take great care to repay society for trusting them enough to give them a good job.

Then again, the old bad boy makes good is usually a good Hollywood story, and, unfortunately, that's the only place that sort of thing happens…unless the great wheel of fortune spins, and someone lucks out big time. The story of someone making it rich on roulette is a real stretch…even if they put $1,000 on a number that wins, they'll only get $33,000 in return…and, if they had a grand to bet, they weren't that bad off to begin with. As old T. Barnum said, the best way to double your money is to fold it in half and put it back in your pocket.

Day 7:
A week of days, a month of weeks…so a millennium of what?

Whole organizations exists that offer services to those in need, and these organizations are often elitist and hard to join. There is a large line between those in need and those who are able to help the needy.

It is a noble and rewarding act to help someone. Perhaps that is why we have such sophistication in charitable circles. Administrators often earn huge salaries for their services. This can raise doubt about the sincerity or motivation behind these people. An entire group of helpers go to school and take social work, psychology and counseling, so, they are in better condition to help the unfortunate. That turns the whole dynamic into a business. The well-trained helpers need people who are less fortunate and rely on their grief to get paid or receive aid. This makes these people protective of their jobs, and highly motivated to exclude volunteers who are willing to step in and help people just for the sake of helping them…all for free, and advice that is often poignant, from experience.

It can be a personally rewarding position. Helping a needy family is always a good deed. Getting a professional salary to do it makes it even more rewarding.

When we are well taken care of, are we willing to turn over what we have to someone who has less? That is a huge question. It is better to give than receive, but if we give everything away to help someone else, we, in turn, become just as needy as they are.

How then, can someone who feels called upon by God to help others actually help others when there is a whole civil service whose jobs and careers depend on having needy people to help? Helping the homeless has become a jealousy guarded position that people only say needs people to help, but then deliberately screens out people who want to help because it is someone's job to help, and they don't want to lose or share their job with other people.

The losers here are the poor and needy people who must have support and handouts to live. They don't care whether a volunteer helps them, a government clerk, or a well-paid foundation member. They just want the help. If that help comes from someone that has been in their position and understands their pain and suffering, they would feel proud that there is no distinction between them and the upper class they rely on for support. It would help them understand that humanity is on an equal footing, and that circumstances are often the only difference between who asks for help and who offers help.

The question of motives is always important. Is the desperation of a spoiled rich kid, a selfish brat that suddenly sees what a greedy, self-serving, gluttonous pig he has become, with a feverish desire that fits with his desire to always get his own way, the reason to help others? Is it redemption for years of selfish abandon? Or does an understanding or realization that giving and helping others can somehow make up for a lack of love in life, or perhaps find solace for personal pain and loneliness?

Do I realize that I don't want the special treatment I receive, that I need to suffer to learn, I need to be deprived to appreciate the bounty I enjoy. I would like others to know what I have received, to have what I have had. Or to have what I have not had.

My arrogance placed me in an isolated situation, where I was an outcast, a person that others saw as depraved, and a disgusting scraping off their shoe, detestable and repugnant.

Have I become a greedy, selfish human child, someone that always has to have his own way, receive instant gratification, someone that has lost patience and the ability to put others before him?

I wouldn't like someone like that, and I could never live with myself if that was my destiny. I would have to change, and with my history, I would want the change to happen as soon as possible.

When you realize your life is a horrible mistake, how can you atone for your sins? The Bible is the source of all spiritual knowledge, and it states that Jesus was sent into this world to die for our sins, his perfection killed by sinners to save sinners. Only through faith in his message will we ever come close to forgiveness for our sins. We are consummate sinners, and we must learn to live a life of grace, as close as we can come to grace in this world. Jesus gave us an example of perfect humility, a life we can learn from by careful study and worship. We must look forward to forgiveness, and not look back at our sinful ways, or those sinful thoughts will return and we relive the wickedness that is our nature. We must fight the good fight; strive to uphold what is blessed and pure, seek to find what is good and clean. Only then will we learn to live with our faults; only then will we ever have the strength and purity of heart to tackle evil and wickedness. We must trust Jesus to bring us closer to God. Our God is a God of love, of peace and forgiveness. This should be stamped on our hearts; written on our souls; scratched on our hands; engraved in us and hung around our neck. It is the only hope of deliverance we could ever hope for. Living the good life is what we are supposed to do.

We do not receive rewards for it, we receive God's love, the greatest gift we could ever imagine. Wisdom more valuable than the purest of gold, more exquisite than the brightest of precious gems, greater than all that can be found in this wonderful creation we are privileged to share. We need to look after this world; we are trustees for God, stewards of the Earth, caretakers of this paradise he created for us. We must thank him, praise him, worship him, and give offerings of ourselves to Him for the many great things he has given us out of love. God loves us more than we can ever understand. His patience is like an endless river, but his anger is like a violent volcano. He is slow to anger, but quick to love and understand. Through the blessed love of Jesus, through the selfless sacrifice he made for us, we are able to show our Father how grateful we are. This cannot wait. It doesn't have a best before date. It is something we should offer freely and freely offer when we understand what it is that He might want from us. Anything and everything we have is His, we should be thankful that we could give, and be joyous that God cares about us. We are unworthy and we are ungrateful and insolent. God's immeasurable love is beyond our understanding; his love is what holds our world together.

His presence courses through this universe, underlies all matter, and structures that which has no form. He calls into existence that which doesn't exist…hard to imagine, but that's God, He's able to do anything. From emptiness, He brings substance; from nothingness, he creates existence; from chaos, he brings order. God is within and without, the beginning and the end, the alpha and the omega. He was there before us and will rule beyond what we can fathom. He is above us and beside us. We are God's children and we must learn to obey his wishes. Only through obedience could we ever find favor; giving our love and finding what favor we can with God is the very essence of our mortality.

Life without God is not life. Remember what Cain said when God sent him away after killing Able. It is a way to darkness and despair, the way of the devil and the curse of evil. Only goodness and mercy will remain, for God will always be, and he will always be our God. Through his son and generosity, humanity is privileged to worship Him. Praise Him whom through all things flows. He is the meaning to confusion; he is a God of understanding, not a God of turmoil, a God of stability, not disorder.

The meaning of your life can be discovered through personal honesty. No matter how successful or rich you are, there are more important achievements in life. These are based on quality, not quantity. They include all the character traits everyone thinks they have, but are only present because you are the one doing the assessment. When you assess yourself, it's easy to gloss over your imperfections and paint yourself in the best possible light you can. No one wants to admit they are selfish, unkind, and impudent, insolent, arrogant and sometimes even cruel. People often act without thinking; their reactions are indicative of their true feelings. They curse, lash out, and respond out of selfishness. Upon further reflection, they realize they acted hastily and out of anger, but the immediate damage is done. Kindness is a reaction just as much as anger is a response. Some people naturally respond with patience and respect; it's in their nature to do what is best.
They are the people that understand what Jesus was trying to tell us. If we learn how to respond with love, most of the problems in this world would vanish.

Life is full of lessons; whether you learn from them makes all the difference in the world.
Success is finding favor with God. Only then will your labors be worthy.

I had everything, but traded it for nothing.
It seems my ship has come in, but I still need to swim to get aboard.
Learn for the glory of God. Learn to work for the glory of God and leap for joy!
Rewards are returns on good deeds, not good intentions.

God rewards people for righteous living; people shouldn't expect rewards but live righteously because it is right.

Day 8:

Eight days a week, I love, I cry, and I ache, dum Te dum (chorus).

I feel trapped in a box of my own creation. Think outside the box. Believe in what you write and write what you believe. I write therefore I am? (Descartes would be proud).

It was a lazy rain. Occasional cloudbursts added to a drifting mist that engulfed the city in a foggy blanket. (switch to the book, and finish that last chapter…)

An occasional surge would remind you the clouds were crying, followed by a drifting mist that hung over the landscape and clogged the sky. Possible sentence for a paragraph, or just follow the storyline—don't write outside the story…you just create a new story.

I see a lot of strange things in newspapers, and wonder why people read such crap.

I hear a whole lot of stupid that I had to include in my diary. A public school teacher was arrested at a bozo airport while trying to board an international flight. When searched they found a ruler, a protractor, a compass, a slide-rule, and a calculator. At a press conference around noon, the Attorney General said he believes the man is a member of the Al-Gebra movement. He didn't identify the man, but confirmed he was charged with carrying weapons of *MATH* destruction. Wow, some sophisticated joke, or someone is way beyond stupid…but wait, there's more!

'With Al-Qaeda and Al-Gebra, there is a problem for us, the Attorney General said "We don't understand it, and when we ask, they go off on tangents…and tangents on tangents, always denying solutions by means and extremes, and always go off on geometrically clever tangents in search of some 'Absolute Rule.'

They use secret code names like 'X' and 'Y', and refer to themselves as 'unknowns,' but we have determined that they belong to a common denominator of the axis of medieval with coordinates in every country. As the Greek philosopher Isosceles used to say, "There are 3 sides to every triangle."

The Attorney General sent on to say teaching our children sentient thought processes and equipping them to solve problems is dangerous, and puts our entire government at risk, especially when balancing the budget.

We don't know what this Absolute Value is, but it must be some nefarious solution to the instability of the wide gap between the rich and poor.

We can't have them solving things like that. Our local expert on numerology said their name Al-Gebra is Arabic in nature, and could have to do with algebra, a system to solve mathematical problems.

Wow…if the Attorney-General believes something like this, no wonder the world is screwed up…everyone is stark-raving, blooming insane!

When a lesson is learned, the benefits can be impressive. That can happen if someone actually learns from the lesson, and if someone acts on what he has learned. What happens when someone learns, agrees that it is better, but has a hard time applying the corrections? That person is doomed. A familiar parallel to modern society not learning from all the mistakes of history—get a global agreement to destroy all weapons, lock up the bad guys or put them on an isolated island, and hope goodness wins. Give peace a chance…for real: however, no one would make the first move, beat weapons into ploughshares. They'd be defenseless, and afraid the other side would take advantage of that and wipe them out. Unfortunately, a very real possibility, thanks to our warmongering and violent tendencies. Are we really savages, or are we capable of becoming an advanced and enlightened civilization...something that might make ET make contact. I doubt the rest of the universe is always at war...especially on their own planet. Another planet maybe, but not against themselves...that would be like a group of ants all splitting into factions and warring against each other to capture the Queen. Very "Alice in Wonderland-like," but true. Moreso if you can step back and take a really objective view of our planet, exactly what any alien race would do. No wonder we seem alone in the universe...who'd make friends with people that would stab you in the back, just to get your technology. Yet, Urban Legends have it that the U.S. is secretly working with aliens, just to get little bits of technology...that shows how much we can trust each other, which is zip...nadda, nothing. It's really a miracle we haven't blown ourselves up with out nuclear arsenals, or did ET intervene and stop a few accidents here and there?

If we realize sin is bad, and try not to sin, we can't always become sin-free. It is our nature to sin. We are sinners. How can we live with grace? We know our old ways are wrong, and try to better ourselves, but sometimes, old habits are hard to break. How can we break a nasty habit that we need to get rid of? Pray, read the bible, and associate with good things. That sounds like an easy solution, but it can sometimes be almost impossible to set into motion. We are a society of selfish technology addicts, a group of advanced humans that no longer fear the unknown, and we even attempt to explore areas that were once considered impossible to visit. We now dream of the moon and the stars, and our science is focused on defining and explaining what constitutes our universe.

God wants us to learn, but how much does he want us to learn about his creation? The more we uncover, the more complex and miraculous His creation becomes; the more we discover about ourselves, the more we realize we need to change.

I found out there is a great deal I need to change. The more I change, the more intimate I become with my own psyche, and the more I have to change. It's like an onion. I peel back one layer, and the next layer just invites more exploration.

To simplify things, I hold certain principles to be true. God created us in his own image; he imbued us with intelligence, emotions, and spirit. He breathed life into us. Therefore, we are expected to use his gifts to better ourselves and discover how to live a good life. It might sound easy, but it is a very difficult road for some. Some people are naturally full of common sense and love. They hear the words of Jesus and believe wholeheartedly. Their lives are dedicated to the love and peace Jesus tells us can be ours, provided we live the way his Father wants us to live. Some people have this gift; some people have to learn the hard way.

Why do I always have to learn the hard way? Even when I learn something, I have to go back and re-learn what I just learned. It really makes me feel stupid. It can seem to me that I didn't learn the message the first time, and I have to go back and learn all over again. Perhaps I am too stupid to be a man. I feel like a child. I act like one, talk like one and still think like one. Yet, I am a man. I absorb knowledge, but I don't act upon what I learn. I'm like an intellectual sponge that slurps up data but never changes. I wish I could be different. Perhaps my bad behaviour is a result of habit. I have developed a habit of sinning and doing the wrong thing. I started when I was small, never received correction, and this is the result. I need to start correcting myself. I need to change. I realized this long ago, but here I am making the same mistakes all over again.

I seem to make plans but never act.

God invented irony, as his unknowable plans often turn the very reason for our destruction into the vehicle through which we are ultimately saved. God can even use our failures and weaknesses to turn our lives around. When God

calls, you know it; his presence is everywhere, and his glory shines through whatever flimsy screen we try to erect to hide our sins. God takes our weaknesses and strengthens them, he takes our imperfections, and uses them to show the power of his word. He is made perfect through our imperfection: that takes a lot of wisdom, and power.
Learn don't burn.
God appreciates you for the good things you do, not for the things you thought were good and should do. Act now.
Focus on the good in life and good things will follow. Learn to work for the glory of God and enjoy it! There is no more rewarding experience in life.

True success is knowing you've found favor with God. You don't have to tell him, show others, or keep a record of your good deeds; God knows what you do in secret. Be humble and proud that we are able to understand God enough to worship him, sing praise, and create mighty works in his name.

Day 9:

What does a month of Sundays mean…why no month of Mondays?

Sadly, society doesn't work on the fairness applied formula. Ultimately, people are too caught up in their own problems to stop and listen to yours. Getting someone to care usually takes money, or an outrageous act that stuns him or her into acquiescence. High priced therapists will sit and listen to you, give you a response, but bill you for the hour or session. If you do something crazy, the shock factor doesn't give you a lot of bonus points. It lasts about as long as the interest in a mime—everyone's speechless, for ½ a minute, then they're back into their own world, their own problems, or whatever's on the table for the day. Unfortunately, I describe using this shock factor in my book **Drugstore Cowboys**, and I use it to make every too shocked to believe I would blatantly grab drugs from underneath their very noses. It worked, and I got away. Blast 'em with brazen insanity, and then slip off into the urban shadows…worked every time. If you want to see some incredible examples, grab a copy, and prepare to be amazed, everything is based on a true story, the names changed to protect the guilty, and events altered to perplex police. Having admitted that, you can see how my journey to spirituality had quite a few bumps in the road. I don't cater to television-star evangelists, as I still remember Timmy and Tammy Baker; they always want to preach the right thing, but then they launch into money making drives, like "okay, get out your cheque-books or your credit cards…NOW," and usually have some offer for $20-70 bucks, which would add up to a lot of collection plate cash if they were in a church. I don't know what they do with the money, and that's always a big caution sign for me. I'd rather buy a bunch of good food, and then take it to where it's needed and give it out for free…and make little packages that they can all use: peanut butter, new socks, underwear, toothbrush/paste, brush, razor, gloves scarves, and other things that would help their life on the street.

At least you know the right people are benefiting from the money, and if you do it yourself, your time and labor is free, so that's more money to spend on help the poor. Prepacked bags of food are always helpful, but you have to remember what these people don't spend money on, and that's basic stuff, like socks or underwear.

Sometimes the desire to do well is thwarted by the chance to do good works. It's like you're trying to work someone else's turf, and they get all protective about it...kind of misses the point, but ?
You can't just run up to someone and offer a good deed. You need to find the need and be ready to fulfill the deed. When it's obvious, it's like too many cooks in the broth...everyone tries to help and ends up screwing everything up. Like this time I helped an older lady carry her tote up the Skytrain stairs...and some ditzy idiot tried to help by grabbing it, and ended up overturning it and spilling her lunch. Once she saw the damage she did, she apologized, and beat a hasty retreat. Like, opps, screwed up here, better move on and let them deal with it...right, the unwanted helping hand. But, you stayed behind, helped her pick up her lunch and organize things, then carried it up one side and down the other...she was so thankful she gave me a bus ticket, after I said no over and over, giving it back, until I recall some societies have customs, and rewarding someone is a custom, and refusing the gift would be an insult. Or something like that. So...did that qualify as a good deed, or was it too screwed up.
Finding dog tags and then taking them to the address on the tags qualified, as the guy was really grateful, and a bit surprised I went out of my way to return them. Also, giving the guy my gloves after he mentioned how functional they were, especially when he didn't have any and it was cold. Hmmm...a good deed is supposed to be done in secret, as God sees in secret, so I should do a good deed and then forget about it...yeah...forget about it, with the Bronx/Wise guy accent.

UnunPentium 115. That was a surprise. Bob Lazar mentioned it in 1989—no one had heard of it, and then it was discovered in 2003. So, are UFO tales really true? Well, you've got your own picture, so that should be enough, but the whole UFO thing doesn't seem to fit into the Bible...then again, the Bible is full of UFO's and universal morality guidelines...maybe that's why ET visits us, as we are the chosen people of God? Now that's an argument few could refute, or waste a lot of time bantering back and forth.

So, selfishness for self-fulfillment and a healthy bank account, or total philanthropy, and a desire to help others? One is egoistic, the other loving...hence, free will and choice.

Charity...or love, is patient and kind; love does not envy or boast; it is not arrogant or rude. It doesn't insist on its own way, but bears all things, believes all things, and hopes all things...(1 Cor:13:4). It sounds equitable, but why do people fight over it and deny others the chance to do something, just so they can do it themselves? Like a bunch of hormone-riddled bachelors, trying to pull out a chair for a beautiful woman, or open the door for that pulchritudinous blond?

Why was it harder to get a job as a volunteer than as an employee? That's a fair question, and happens quite often. The volunteer agency started off as a charity, with people working for free, then it turns into a bureaucracy because of all the money donations, and those charitable workers turn into salaried employees that are just as mean as their counterparts in the business world. Add arrogance and self-aggrandizement, and that covers many so-called charities. Money grabs or lotteries...all after the almighty dollar.

Everyone likes to do a good deed...but holding the monopoly on good deeds turns into a bad deed, as they're bursting with selfishness and greed, and their charitable constitutions were left by the curb, and they scratch and bite their way up the corporate ladder worse than the understandable go-getters in the world of money. One screwed up world, extra crispy, coming right up.

William Wilberforce. Now that guy was a true humanitarian. A personal crusade to eliminate slavery that cost him a well-paid government position…that is true charity, but…there aren't too many situations like that anymore: his time period allowed for his **"Amazing Grace"**…good movie.
As Jesus said, "Live to serve," like he showed the Apostles by washing their feet…not too many people pick up on that act, as no one wants to scrub some homeless guy's feet. Then again, most people only focus on Jesus' greatest act…dying for our sins, and making it possible for gluttonous, self-serving buffoons to get into Heaven. They don't get the serving part, but love the Pardon part. Pardon is key to redemption, and not everyone chooses to embrace anything they can do so pardon is part of their future. They actually choose to wallow in sin, picking and hanging on to self-moralization, thinking they know the way to Heaven.

The only way to Heaven is through Jesus. He said: "I am the Way, the Truth, and the Life…all those who wish to see the father must come through me." That is crystal clear, but it is the most misunderstood statement in the Bible. I read and read, and I don't understand…I learn, and I understand, but I can only understand what God wants me to understand…at that moment. Perhaps that's to help my repentance, or help me become a true Christian, but if God thinks it or says it or even hints at it, that's good enough for me. These people that choose personal glorification or the fleeting pleasures of this world must be mentally short on a few screws. How can anyone believe in what they believe…the Bible says, "Be not wise in your own eyes," which is also repeated in other sections. They're smart, in their own mind, and that's always a huge mistake. Only an idiot tells himself he's smart…if someone needs affirmation, then they're dealing with shadows of reality…mists of the mind…things that are not there. Yepper, they're dumb, dumb, and dumber. Wait, you're not supposed to judge anyone. Okay, everyone's the same…we're all smart in our own way, and we all need to read the Bible and make up our own mind.

I remember an argument with an atheist. It didn't last too long, as once he said, "I don't believe in God," I said you're wrong, and walked away. Why argue? Either they get it, or you can share testimony til you're blue in the face. Like you know…he never comes out and says it, but he never agrees with you, and doesn't attend church. Well, you're not exactly there every Sunday either…judge not…judge not.

Pick up the torch, venture out into the world and save starving Africans…right, and get shot by a terrorist or get hijacked by a pirate. The world is not the same place. Evil…we even has famous people using that name…Evel Knievel…well, it rhymes. Marilyn Manson…a combo of Marilyn Monroe and Charles Manson. Well, take a look at the guy that uses it…heck, he even cuts himself on stage. Weird, wacky, and totally inane. Not insane, but inane…a wee bit droll, as the English would say, but definitely not insane. Insanity is what…the belief that doing the same thing, exactly the same way, over and over again, will produce different results. That's insane. That could also be the empirical method of scientists…they always want to repeat results to prove their pet theories…so, does that make them a bit bonkers? Probably, but using that bit of reasoning, we're all totally fried. Too much sun, or too many cosmic rays…yeah, it's got to be the cosmic rays…when the Earth passed through the galactic core on December 21, 2012, we somehow picked up some pixie dust, or cosmic leftovers, and they're taking their toll on us…slowly but surely. Perhaps that happened in 1901, and is the reason for the sudden surge of technology, and not because we found a bunch of UFO wrecks and cannibalized them for spare parts and new technology.

Wait…that's why we have fiber optics and Velcro. ET's don't have shoe-laces or buttons, so they have Velcro on all their garments…perhaps they shared the technology, hoping we'd design a whole new line of clothes using Velcro…when they take over the Earth, they'd have a line of clothes all ready for them. Right.

Trite, right, and then write…off the cuff, no more guff, you're not that tough. Riddles in a rhyme, rhyming riddles, or a Sphinx-like puzzles of mystery.

<u>Writing tips to incorporate into new book.</u>

Chance is a fool's fate. Chance is fate to a fool. A fool's fate is up to chance. Chance: a fool's name for fate. Make sure ideas are in the right spot.

Trim the fat, edit the unnecessary words…savor what works and spice up what doesn't'.

Time pressure. Action must occur with a time frame. Puts pressure on story, gets readers turning pages to see what happens. Ticking bomb or search for Mcguffin.

Turn up the heat. Add more problems, more characters vying for same thing. Lock somewhere—no way out, & problems—let them figure out how to win.

Describe an important item. For instance, mention a shotgun, loaded, sitting by the fire.

Then find a time that it will come in handy. Reader remembers it's just sitting there.

Revision, redraft, rewrite. Revise, rewrite, and edit.

L.I.F.O. – Last in, first out. Jump into action late and get fast. Don't draw it out.

Day 10:

I write, therefore I am.

L'homme au masque de fer (Iron mask)

The Inca Horde, or Mayan God, Mayan Gold:
Short story or novel…

Gold Tailings and Steel Bars

    Wiping his sweat soaked brow on the shoulder of his T-Shirt, Ron redoubled his efforts to pry open the paint encrusted second window to allow much needed air into the stuffy attic. The window finally gave, and he managed to force it halfway. A deliciously cool cross breeze started to fill the attic, disturbing the decades of dust that now whirled about in small eddies highlighted by the slanted sunrays that poured in the front window.

The attic was a no man's land of pushed aside junk, and the hot dusty atmosphere ensured no one ventured up here, so no one knew what was stored here. Mysteries and the unknown always intrigued Ron. The family consensus was that the attic was full of old, totally worthless junk. Most of the trunks, shipping crates and assorted furniture came from the large family's distant and now deceased kin. Since it came from ancestors and far-flung relatives, the current descendants felt it should join the other family heirlooms; hence, it was relegated to this out-of-mind, out-of-sight graveyard. Expensive antiques and liquid wealth went through the regular channels, but personal possessions always ended up here.

There was something tawdry about throwing out a family member's private belongings: it just wasn't good taste; so stuffing it in the attic and forgetting about it seemed like a good compromise. They were destined to sit here and collect dust, until a bored and curious family member had the time to rummage through them. The Ross mansion had been in the family for generations, and had lots of room to store items not suitable for display, but too valuable to toss in the trash bin. The contents were from many historical eras, and only someone with a love of history and family lore would spend the time to sift through the growing collection. This was perfect for Ron Battle, one of Boston's most eligible bachelors, an outstanding scholar and true lover of history, now excitedly rummaging through the dribs and drabs of distant relatives; once wealthy travelers, hobby archeologists, and eccentric collectors.

The attic's humidity was pushed aside by the refreshing draft, and what dust didn't make it out the opened windows was now blown about like small white tornadoes of debris. Ron let the breeze rumple his hair and cool his face as he returned to the open chest that contained his Great-Grandfather's, Aral Battle, last will and testament and long-forgotten personal possessions. Since he only owned a few acres of wilderness in what was now British Columbia, no one seemed inclined to do anything about the land or what his long-lost lost kin had left to the world. The Boston Battles had made a fortune during Aral Battle's time in cotton, tobacco and, before the Civil War, in slaves. No one cared about a few measly acres in the middle of nowhere. The deed was there, so Ron assumed it was still family property. Something worth looking into.

There were also some antique mining equipment from the 1880's, old mining claims, a bundle of colorful old stock certificates, a complete assayer's traveling kit, some personal grooming effects and a compact desk, a very popular item with traveler's during the Victorian Age. The surface was approximately 3 ½' X 2, and about a foot deep, and the entire piece was expertly crafted and beautifully inlaid with exotic woods. The writing surface had a leather cover and swung open on well-oiled brass hinges, like an old fashioned, flip-lid school desk. The interior contained several small drawers, and sectioned off compartments for writing paper and instruments. There were some antique fountain pens with gold trim, a stack of faded stationery bearing the family crest, several odds and ends he couldn't identify, a stack of old letters, a sturdy engineers field notebook and a stack of old shipping labels. He opened one of the small drawers and discovered a fragile looking pouch with a thick drawstring. He opened the pouch and poured an assortment of rock samples into his palm.

He looked at them closely, and then returned them to the pouch when he couldn't find any obvious signs of gold. Ron wondered if anyone had bothered to look in here after the items were returned to the family mansion in 1885, after his ancestor was murdered in the Canadian Northwest. He noticed the inside seemed to be smaller than the exterior indicated, so he looked for some trigger mechanism that might hide a false bottom. At the very front, he found a perfectly inset wooden ring that could be accessed by a small notch in front. A masterpiece of carpentry, the latch was made of the same wood as the interior, perfectly smooth, and impossible to see. If he hadn't been looking for some kind of trigger, he would have missed it entirely. He picked the pull ring up with his fingernail and removed the top tray.

The hidden compartment contained more letters, another notebook, a book-sized rough wooden box, a silk sac, and cord, and a single piece of paper folded in half. After the excitement of finding the hiding place, Ron felt let down. There were no stacks of money, chunks of pure gold, or anything remotely interesting. The contents were almost the same as the original drawer. As he picked up the single piece of paper, the sun went behind a cloud, leaving the attic in shadows. He brought the paper over to one of the windows for a better look, and immediately recognized his Great

Grandfather's neat and meticulous penmanship. The very first line got his attention.

It began: "To whoever finds this letter. Since you are reading this, I am dead. Murdered by Sam Locklear, the owner of the Five Star Salon." His interest piqued, Ron sat on an old chest and read the rest. The letter was shocking. Not only did it name the murderer, it described an incredible find that was apparently lost to history.

He rushed back to the chest and examined the other concealed items. He opened the silk sachet and poured the contents into his hand. The dull gleam of old gold gave his hand a rich steady glow. From the weight of the items, he could tell they were solid gold. There were some heavy rings, thick sections of chain, a couple of smooth nuggets, and an incongruous ancient amulet, with intricately carved marks that looked like the symbols carved on Aztec temples. Ron put down the other gold and took the amulet over to the window to study it closely. It was definitely Aztec or Mayan craftsmanship. Ron had studied pre-Columbian art at Stanford, and had trekked across the Yucatan peninsula to see the massive temples that had once been the centre of South American Indian culture. There was no mistaking the style and exquisite skill that went into creating the amulet. At the top of the trapezoid shaped artifact was a depiction of the sun; thinly etched beams shone out from the centre, and on the bottom were distinctly drawn Aztec pictoglyphs. What really puzzled Ron was how or why this object was in a sack of miner's gold supposedly collected over a thousand miles away? As far as he knew, his Great Grandfather never visited Mexico, and even if he did, how did he get hold of a precious work of art that belonged in a museum? Clasping his hand around the object, he returned to the chest and picked up the notebook, hoping it might shed some light on this peculiar turn of events.

Looking around for a comfortable spot to read, he saw an old sheet covered easy chair, and decided it was perfect. Ron took the book and sheaf of letters over, removed the chair's cover, and settled down to read. He hoped the answer to the Mayan or possibly Olmec artifact was in the old papers: an amulet wasn't the sort of thing you just stuck in a bag and stashed away in a hidden drawer. There was a story here, and the only link to the tale was the letters and the notebook. Ron opened the notebook and flipped through pages. The handwriting was easy to read, although written in a stylish and flamboyant manner, with specific dates preceding each entry. Glancing through the stack of letters, he noticed the majority were from well-known Universities, Academic Associations, and various research institutes. He made himself comfortable, and eagerly began to read.

(see Olmec God, Olmec Gold—novelette, still in editorial hands…why?)

God has a deep love for us that transcends normal emotions. He demonstrated this to us by allowing His Son to take human form: to live with us, to feel as we feel, to suffer as we suffer, and then offer us salvation from the sorrows of this world. This is part of God's master plan for humanity, a chance for us to escape the evils of this world and embrace what is pure, clean, and pleasing to God. We are so far removed from the grace and holiness of God in our present form that it took the sacrifice of our Lord, Jesus Christ, to atone for the sin into which we are born. We might not understand the grand scheme of things, but we must learn that we must reject sin and embrace God's word for true peace and happiness.

This new and final covenant with us allows us to seek the grace of God and know the forgiveness of God and his love through the Holy Spirit, a miraculous form of pure love and holiness that is part of God's presence. Jesus fulfilled God's purpose. He brought us the gift of God's words, a gift of love so profound and so deep we are stunned by its power and purity.

Jesus brought us the keys to Heaven. Whether we learn how is used is up to each individual. The world has many snares, many pitfalls, and many illusions. We are forced to make a choice. We can be tricked by worldly sensations into believing the lies and philosophies of those who are helplessly in love with their wealth and power, unable to look beyond their possessions to a greater world of love, thereby ignoring the message of God to live by a different set of rules. God's message is a message of love; love thy neighbor as one might care about oneself. To think about the needs of another person rather than selfishly insist on getting what you need, demonstrates an awareness of life beyond a personal level. Recognizing the needs of others might be more important than your own shows compassion and care towards your fellow man. This is the sort of selfless love Jesus told us about; that putting others ahead of yourself is the first step towards humility and awareness.

Altruistic behavior is part of personal salvation; when you reduce your own self-importance, you grow in morals and sympathy.

Day 11:

Another day—witty comebacks always come to you after you need them.

Things We Can All Relate To, Understand, or Empathize With: Passions of the Ego, Acts of the Id, and Reflections by the Superego

A few things that can get your down, but through recognition, you could find redemption.
If any of these things could be you on occasion, admit them, confess them, and ask forgiveness. If your attitude is one of repentance, you will be on your way to acceptance.
This is the not nice list:
Jealous, Envious, Covetous, Disrespectful, Arrogant, Gluttonous, Ignorant, Careless, Blind, Mean, Demanding, Lackadaisical, Haughty, Irresponsible, Untrustworthy, Indolent, Greedy—me not we. Angry, Liar, Inconsiderate, Judgmental, Conceited, Stubborn, Sneaky, Pouting, Ungrateful, Cowardly, Insolent, Reprehensible, Shameful, Immature, Atrocious, Dishonorable, Appalling, Self-important, Contemptible, Foolish, Hateful, Rude, Demanding, Devious, Delusions of grandeur, Unforgiving, yet lusting after forgiveness. All qualities no one likes in others, and should avoid in themselves, but that is often not the case, as these qualities exist on various levels, and some people are okay with a tiny bit of each, thinking it is a "white" quality, unimportant, as a "white" lie.

You can read the bible with gusto and find love and everything you cry out for; tears fall from your eyes as you read about true love, friendship, and brotherhood, all displayed by the great and faithful Saints of Jesus. You know your heart is full of love and you don't want sin in your life, so why don't you follow the path of Jesus and do everything that is written down. You may scrupulously examine your behavior because you do not want to sin, but when you honestly examine every facet of your being, you can still make a list like the one above. I might not be guilty of everything, but even the thought of something is a sin, and therefore something to avoid. It is shameful and I would do anything to erase that stain from my being as it is not who I know I am. I am caring, I am generous, I don't steal; I've been a terrible person, but unless I am granted forgiveness, I am doomed. But if I die for what I've stupidly done in the past, I will die doing what is good, for I no longer want sin in my life. Well, that's the way it's supposed to work.

I beg forgiveness, but can I forgive myself? I still see the awful man that acted like an idiot, and hate that man. If I am to be judged, I will nevertheless be judged by the standards I use to judge myself, as I have read enough of the Bible to know enough to fear God. I am scared because I was so terrible, and I am still scared because I know the ultimate power of our Lord, and the power of our Father, who made this world. Only the Father can make something out of nothing, and only the Father can forgive you. If you have real faith, that's when you begin to believe in forgiveness, and feel better.

I read that our Father is a God of mercy, but I also read that he has a terrible wrath for sin. I also hate sin, and when I see it around me everyday, I know I want to be a good person, not a bad person. I have to have faith that Jesus really does love me, someone who is lonely and mentally convinced there is nothing about me that people would like, let alone love. I know I have love inside me, as I weep and cry out when I read the message of peace and a loving, responsible society. I have never been accepted into this world, and it shuts me out and there is nothing I can do about that, for I am the one who disobeyed their commands and laws. I paid the price they wanted for my crimes, but all my other behavior is something that will never be forgiven by the shortsighted people of this world. Jesus said if you love this world and the world loves you, you do not have your eyes on Heaven. If I could find love and true forgiveness, I would explode with happiness. I would endure anything to find that, and the Bible does say that suffering builds endurance, and endurance builds character. Character builds faith, and faith gives us the hope we need to understand the messages of Jesus Christ. Christ, because of love, to Earth as a mortal man, when he had the power to be the greatest King this world would ever see. He gave us the true word of God our Father, and was killed by jealous and sinful men who didn't accept him as the messiah, for it meant they would no longer be exhaled among men, but become humble and respectful of everyone. They loved the transient power of this world more than the everlasting life and happiness Jesus offered them.

God knew this and in his magnificent plan and love for righteous humans, made a new covenant with us, including the laws of Moses, a new covenant that includes all mankind, the gentiles, so all may worship him, trust in him and love him. All men are now equal in his eyes, and we all fall short of the glory of God. Some are blessed for their fantastic works for our Lord, but all men can be saved and blessed through the sacrifice of our Lord and Savior, Jesus Christ. We must suffer, for our Lord suffered and could enrich our faith knowing our Lord suffered, just so our sins could die with him, and only out of love for us. That's a lot to ask someone to endure for people that ended up mocking him, spitting on him and then nailing him to a tree. He died not for the righteous, for they were not in need of such a sacrifice, but for the sinners and unrighteous of this world, that they might be saved by redemption from hearing his words.

The Bible has taught me this: all men must share in the love for our Father, love each other, for love is the glue that binds the righteous together and gives joy to life. We must live with each other, but it is not as easy as it sounds. Men love the fleeting pleasures of this world, sin, sexually immorality, wickedness, hate and the love of power and money; things that separate them from the Holy, Holy realm of God, for He is pure and unblemished, untarnished by sin and too exalted to listen to dirt, blaspheme and the foul boasting that comes from the unclean mouths of unrighteous sinners.

I don't know if I am able to write these things because I have conquered internal emotional pain, for I lack understanding in the ways of the Lord. Being spiritually healthy is preferable to being burdened with guilt. I know I need to follow the word of the Lord and only do what he would want me to do. I've learned that from the Bible, yet I know many religions teach a similar goal. If everything is a choice, I believe I've made the correct one.

Devoting my life to the Lord would be an honor I have a hard time believing I could ever fulfill. I know the Lord is all-powerful and is able to do all things. I need to cleanse myself and find the faith and total devotion my heart yearns for, as I always seem to do things with all my heart. I cannot find forgiveness from this arrogant and judgmental world, and believe with all my heart that I am truly sorry for my past sins. I am unworthy of many things, and would endure all things for forgiveness. I am a sinner and find only sorrow and pain in this world. In my stupidity and weakness, I fall prey to temptation that drugs bring, as it has been a huge mistake of mine to think that they take away my pain. I know Jesus would fill me with joy and take away my pain, but in my weakness, I fall prey to momentary hurts and turn to the only source of help I ever knew.

This is idiotic on my part, and I know that the joy I once knew would be given to me tenfold. My low esteem always tells me I do not deserve this, as I have been told by too many people, too many times, that I am unworthy. Never listen to people, they always like to build themselves up by putting others down (Wow, lucky I'm not like him). Christ mentioned judging others was a bad thing—it's best to love others, and do all you can to help. I sinned against the Lord and destroyed the faith I had in myself to live a righteous life. If this is the case, I would rather live a good life, expect no rewards, and hope that the grace of God would have mercy on my foolish behavior. It's amazing what you can learn with an open mind and a good book.

I need to pursue Him with a sober mind and sinless life. I'm writing this to myself so I can try to understand what is wrong with me. I have to find strength, endure my suffering, and use what I've learned as best I can. I am what God made me. Whatever path I've taken, I want him beside me. If I can be worthy, I'll follow him to the ends of the Earth. Sitting here, without the demons screaming inside me, everything seems calm and peaceful, everything seems possible. Then I wake up and suddenly think I can't exist without instant relief. I know what needs to be done, I just have a hard time doing it. I don't know if I'm possessed by a certain demon, but if I am, I ask the Lord to cast him out. I'm writing this as a personal message and a private prayer to Him, so that he knows what is going on in my mind.

I don't know anything about the spiritual world, but I think the Lord knows what I'm going through, as he is omniscient and all-powerful. I honestly want to serve him, but I've become terrible at communicating with people, and have been turned away after offering my services. On a personal level, that hurt me, but after what I've read about the sorrows Paul and Peter and the other Holy brothers went through for Jesus, I must be a craven puff of wind if I can let that keep me from doing God's work.

God's work; it has a special ring to it that elevates it to a joyous task, no matter what difficulties you face while doing your duty. You are doing something that matters for the only thing that matters, our Heavenly Father. There are many already doing His work, people that are famous, draw huge crowds, have budgets bigger than a small nation, so what could I do that would make a difference and bring the Holy word of God to someone that has never heard of redemption and salvation. I understand a lot about a part of our society no one wants anything to do with. Jesus said to associate with all people; he ate with tax collectors and sinners, healed prostitutes and was crucified with thieves. I could minister to these people for I speak their language; whatsoever that might be, if it turns a sinner into a righteous person, there is rejoicing and happiness in Heaven.

Day 12:

As the page turns.

Jesus said that he came into this world not to judge it, but to save it. He came to heal sinners, for when asked about being associated with the lowly, "Those who are well have no need for a physician, but only those who are sick. I came not to call the righteous, but sinners." (Mark 2:17)

When you want to witness for our Lord, there are stumbling blocks, for there is always someone else in position to do Heavenly work, and those people are unwilling to share, step aside, or listen to what someone else might have to say, even after what they have been repeating for many years has grown stale, and make no difference. Sadly, they focus on the collection plate, and not on the people, the ones they're supposed to save. This isn't what the Bible says, but that's the human condition for you, and the human need for power and recognition. It seems universal, and even those who are aware of what the Bible teaches don't always practice what they preach. Bunch of hypocrites...yet I forgive them, they're human.

In Revelations, Jesus told John how disappointed he was with the major churches of the day. Just because it's a church doesn't mean there isn't selfishness, arrogance, and the desire for power. Jesus talked extensively about those who loved the adoration they received from man...suggesting any notice from God outweighed everything beyond understanding. Jesus also told us he came to serve and not be served, and that we should adopt this sort of mindset. If we are prepared to be nothing, and humbly accept that deep down, God will exalt you and give your honor. I don't want to stereotype, but I can imagine that a lot of the old slaves were truly righteous, if they served with the Lord in their heart. I won't go on about the slave owners, as that is such a despicable act I really view it with abhorrence. Who could ever have the audacity to think they could own another human being? And all the time they were keeping slaves, the constitution stated that men are free, and able to pursue life, liberty, and the pursuit of happiness. The hypocrisy is so obvious, it's unbelievable. William Wilberforce ended slavery in Britain, yet those cotton picking states all wanted to get their cotton picked for free…hence the Cotton Ginny. Inventions aside, they still had a real attitude towards another human being…this went on into the 60's…back of the bus, separate water fountains, and even white only swimming pools. I'll have to stop talking about this, as it really gets my blood boiling…I think everyone is special, and only time and circumstance change who gets the big toys. On top of all this, I felt this way before I picked up a Bible; being a slave always seemed so unfair to me, an attitude I clearly remember I hated as a child. I really couldn't understand how another person could harbor such hate towards another person…whipping them, depriving them of food, and all the other cruel things they could do…wow. I guess the innocence of a child is sometimes a very good thing, something we shouldn't just toss aside as we get older. Jesus mentioned being born again, so that seems like re-entering the pristine state of childhood, with no hatred, and only love. Well, not all kids were like that…there were a lot of bullies around, and I really disliked them. Love conquers all. That, above all else, is truly rewarding.

I volunteered to help at a Church gift shop…a shop that resold donations at reasonable rates. Sadly, the shop was merely in a church, as Church values were not considered or put into practice. I found other volunteers all had their tiny area of expertise, their own little fiefdom where they were in charge; the whole structure of the store was such that people tried to outdo each other by being more important than each other. This is entirely against what Christ taught, and I was finally so disgusted by it I had to quit. I originally thought I would be helping and serving God by donating my time…it was a particularly bad time in my life when I felt worthless and discarded…volunteering to help others was the only way I thought I could lower my self enough to understand I could only go up from there. It might sound strange, but when you reduce yourself to nothing, you realize you really are nothing, and you can then become a true child of God…something that will raise your spirits beyond what you ever considered happiness to be.

The body of the Church has many members; each has a gift, and should use it as best he can, but when the members of the Church are so great in number, many who want to help are unable to help because there are "experts" already in place, members who have diligently done their jobs for many years and yet have not made a real difference. It seems jobs and positions have become more important to the people who hold those jobs, even when another member might reach sinners with the Word; for in the beginning was the Word, and the Word was with God, and the Word was God. We must all work to this end, each doing what he was called to do, what he was best able to do.

Day 13:
Question everything, but accept all but the truly evil.

You could be a writer...but only if you write. There are so many writers out there that have legal training, doctorates, and degrees in history/science/English, or whatever, people who understand grammar more than you...wait a minute, no one understands grammar, as grammar has no real rules, especially when you're creative. Look at T.S. Elliot: The Wasteland. Or James Joyce. The only reason you might get published is because you have absorbed enough examples of writing that your own pathetic attempts have been elevated by constant encouragement from classical authors. Your poetry might mean your prose is adequate. Mapping out a complex plot requires a clear head, large brain, and the ability to think many moves in the future.

Some people are professors of bullshit; that could elevate you like an astronaut.
At least you have some professional rejection letters...make a collection, and paper your walls.

Make sure you have plans for every day, unless you are working. Even simple plans such as going for a walk at a certain time, going shopping, and library research...etc.
It is important to build some kind of structured time. Jobs only give you specific times to be somewhere and do something...make sure your free time is full. Appointments have to be made up if there are no immediate necessary meetings.

You need structure. Commitment. Anything is better than nothing. Just by saying you have a walk scheduled for 11:00 gives you something to build around. It may not sound important, but it is for you.

Remember the book you were going to write about hitting rock bottom and finding "do's," things to do that will give you a sense of accomplishment. Many people fall apart because they never have anything to "do." They get out of detox, and end up back on drugs, mainly because they don't know what to do with themselves. Boredom is a killer...try to be busy doing something for a purpose; accordingly, that gives you a purpose, and helps you build a proper life. Always have something to do, and you will always have a purpose and know you are needed.

Things I noted that I could do, as an example:
1: Find the research you did on that book.
2: Organize your writing into a sensible collection of stories, research, and exercises.
3: Take walks
4: Clean the house
5: Take the bus downtown or take a bus trip somewhere.
6. Watch others. Have a coffee and read a book.
6: Remember what authority, discipline, and structure can do for you.
Save your money...what little you have. You got by with nothing last month, so save what you have and get by with nothing again. Try and sell stuff. Typewriter, art, anything you can make a buck from.

It is always important to break out of a rut. Unless you can change your behaviour and current habits, you will never change.
Remember you want to live and help others. You have to help yourself right now. No one else can help you. Reach down into the bowels of your inner

self and pull up the strength you need to fight. You can defeat it if you focus yourself. Remind yourself about what you have conquered before. (school-university, jobs from nowhere, happiness from consideration, peace from confusion, life from desolation, nerve and audacity to do what others were afraid to)—video player for John, recovery from accidents, withdrawal from heavy drug use, the A.R.F. recovery period and the possibility of losing your right arm, standing up against Red, grabbing cough syrup and running out in daylight, showing up and just talking your way through things. Remember driving the Buick home in reverse…and driving the MGB-GT using down-shifting and the parking break to stop. Wild ride, but who made sure I got home? Remember all the close calls you had, and ask yourself honestly, why you are still here. Overdosed, two in the morning, stretched out on the main CPR tracks to Union Station…and a track walker found you and got you to the hospital in time to pump your stomach…even the doctor visited later and said it was a miracle. So, miracles happen…accept them and be gracious. You had big balls…use them.

You have fought before, and you are still here because your nerve and strength got you through tough times. You survived alone against a hoard of bad guys. You are free now, not locked up and forced to live with bad people. Find good people and learn to make friends.
Try bridge and lawn bowling, if you can slow down and learn to control what controls you. Or, give up what controls you because you want something better in life.

Thank God that you have a chance to live the way you want to, the way you know you should, that chunk of goodness buried deep inside. Fight the little demons that drive you to do what you know is wrong, or what you feel is wrong.

Remember the love you felt when you were young and were lucky enough to feel God's love…not many are that privileged. He choose you for something. You were lucky enough to know God at an early age…stop rebelling and sinning and do something that would finally make God pleased. Perhaps the Holy Spirit would dwell in a clean and sinless me. You are so lucky to have felt what you did…don't slap him in the face…he died for you, so you can live a good life. Wake up. Are you trying to get everyone to hate you? Well, you're succeeding. Remember, if you're losing more friends than you're making, pretty soon you'll have no friends at all. Turn your back on God and you lose. He will never turn His back on you, because he can't deny Himself…He is love personified. Would you ever want to lose that love? You know the answer is no…so, wake up now and get your act together.

Look at your situation as a huge lucky chance to start living life the way you were meant to. Practice what you've learned from the Bible. Remember Pioneer Camp and the good examples you met there. Focus on washing away the stains of sin and evil, develop good memories and helpful things to do.

Day 14:
Is March better than September, just because it has an extra day?

You messed up a lifetime. You will never get that time back. If you want to make a difference, live every moment of the rest of your life for the glory of God and respect his laws and promises.

Laws, as best as you may understand them, and pray for guidance and enlightenment. If you are to help people, try and place yourself in the right position and trust that God will open the doors so you can spread His love and kindness.

Advice has many forms; Godly advice can perform miracles if dispenses from a loving and believing heart.

Life will not change overnight, but if you start on the path of recovery, every step can take you away from the pit of despair you fell into. Don't live in that pit…crawl if you have to, but get out of it somehow. There are people who need your help. You have learned what not to do…help someone resist the temptations you succumbed to and build the life you never had.

Don't think of what you need, think of what you can give others and make sure you live the way you know you should. The most important lesson you need to take to heart is the necessity of being busy with something that is both productive and satisfying.

What you're doing is building a life from scratch. There are a number of steps you need to take.

Perhaps that book you thought about would help you personally while you create it. Sort of a self-catharsis process that lets you learn while doing something.

If you seriously want to write something, you need to lose the drugs. Drugs fog your mind and plug the creative juices you have. You may not be a great artist, but you know how to create things. Create something that you are proud of, something that will stand up to scrutiny by the outside world. Embrace the challenge. You used to rise to the occasion. Dig deep and get the determination, the stamina, the guts, the desire, willingness, keenness, passion, zest, and interest you had. It will return without drugs to stimulate it. You are in a very challenging situation. Imagine the joy you would feel if you could break out of this and find a rewarding lifestyle that brings happiness and success. Don't think about futility. You are here today, so create today what was not here yesterday, something that will remain tomorrow. Be thankful you are here and have the chance to really change. Remember all the drug treatment you went through but never learned from.

Tips on Depression:

The tree story: A man and his son felled a tree in winter. After it fell, the son noticed a green ring, and realized it wasn't dead. The once green tree was now only a stump for chopping and firewood for the stove. They lamented all the years it took to grow, and thought of the shade it provided and the shelter and nests for birds. The father concluded, "Let this be a lesson…never cut down a tree in winter." Like the tree, many things may seem to be dead but are merely unfinished. Finish what you start, lest someone who comes after you sees it as a dead effort and nullifies your work by breaking it apart into more useful pieces. This probably plays out all over the world, with sundry different tasks, all put aside and considered dead and forgotten.

Do things that you once loved to do. Remember past interests and find out if you're still interested. That's where hobbies originate. Or, sign up for a course and learn something new…there are always challenges out there, but you need to look for them. Facing a challenge is the quickest way to discover something positive about what makes you tick. Achieve goals one at a time, and don't go for the really big one all at once…break it down into small steps…things you can do and things that add up to what you want.

Always have interest in others. Caring about things and people. A heart to love and a willingness to forgive. (Forget? working on that one).

Remember you have authority over yourself. Learn to push yourself and finish what you start. If you leave something unfinished, who do you expect to come along and finish what you've started…oh, the dishwashing fairies, or those grass cutting dudes. What about the homebuilding guys…just buy some land, and "Poof" you'll have a house. Right.

Without things to do with your free time is what gives groups like Heaven's Gate a chance to flourish. The old adage about idle hands and the devil's work comes to mind.

The shake 'n bake hodge-podge of Christian doctrine, stuff everyone has heard about, mixed with the science of evolutionary advancement, along with a good dose of science fiction gives a New Age faith something fresh to offer, and people are always looking for something to do that will really make a difference in their lives.

Look at all the stuff you've tried, although you kept your sense of sanity, perhaps a very valuable life-line to true spirituality and sensibility…as in sense and sensibility, but beyond Jane Austen's version.

That "Four Agreements" book was insightful.

Be impeccable with your word; say what you mean, always be truthful, and to the point.

Don't take anything anyone says personally, as nothing others do is because of you.

Don't make assumption, you usually don't have all the facts, so errors happen that persist. You avoid misunderstandings, as people often think of something detrimental.

Always do your best, and you don't have to worry about going back to fix things.

Day 15:

Creation is a dream with work involved.

God had long ago decided a nice, friendly race of pets would be an excellent idea. So, first came the cage (universe), and then came the mice (us), (Adam and Eve, Noah), along with the cheese. Sin.
But God being omnipotent, also knew he would like to have to have pets that understood his world and were able. And pets that were worth keeping. Who wants a pet that might want to eat you? Or be smart enough, or stupid enough, to try and ignore you and take over the really nice cage you made for him.

That sounds like the kind of bonehead or egotistical play that
Mankind would pull, you know, include a lot of needless suffering because they were having a good time here. They didn't realize the owner thought they might be a wee bit grateful for the wonderfully complex universe, this magical maze that this great magical king created

How could the pet have the respect due to the cagemaker unless they could appreciate the complexity and wonder of their cage? And understand the wonder and respect they owed the Cagemaker?
You are the key Lord.
You have always been.
Throughout time and beyond and that much more…
Your Greatness exceeds our covenant with you…(obligation)

A singularly crazy take on what might be, or what might be to be.
God planned to create a creature
Excuse me, not our Lord and Savior, but a totally fictitious Savior.

Once upon a time, there was a great and benevolent King with great and wonderful powers that ruled a fair haven of like great like-minded and almost equally powerful beings; everything was sheer bliss.

Suddenly, decent scurried around the kingdom. Unleashed by a jealous and powerful Prince of His realm, God smiled upon him, as he knew his opponent had absolutely no chance of ever usurping power from Him.

The unruly price quarreled with the benevolent King, forcing the King to banish him from his Kingdom, forever.

The arrogant and vengeful prince was chucked out of the great Kings realm, taking up residence in a lower and less desirable portion of space.

The kings had knowledge of space and time, omnipotent knowledge that would allow them to create, occupy, and inhabit dimensions beyond the norm.

The angry prince sought revenge against the kindly and power King.

The King had created a glorious section of time and space that he was told was to never visit.

The sneaky and disrespectful prince stole into the great Kings new area of nothingness, returning with a heavy infatuation of the new space made by the ruling great King. .

Banished from a …

The heavily divided powerhouse of lesser being ruled by the great King accepted his design for a wonderful occupation of space.

These being were so powerful they could initialize any gravitational or magnetic field they would encounter. And God made the cosmos with 12 dimensions.

Gracious rule and mercy through love.

Planned to expand his creation, but his creation expanded all by itself.

More and more sounds, cries and pleads go up to Heaven, and God listens and rejoices.

God threw down the gauntlet, admonishing the smarmy prince to remember his power and the normal respect the greatest power of all deserved.

It was at that moment that the great King thought creating a race of warm and loving units of energy was a great energy tool, especially if he could get them to go beyond the sum total of their creation. They would have knowledge of the vast and limitless cage, and also learn respect for the cagemaker. And worship Him as God, and their father... divine greatness. Love bears all things, and is the glue that hold their divine life-force together...their soul survives and continues to be an energetic spark.

But, what of the other people...the arrogant, the self-indulged, and the self-anointed Kings and selfish superiority cases with delusions of grandeur—humble them.

No matter, my children, you are watched.

The wise and omnipotent great King instantly solved the problem.

The great King dealt with the problem of the unruly prince quickly. He created a hyper-intelligent race of energy units he would call human beings. They would give him companionship, gratitude, respect, incredible debt, and anything that was good and worthwhile.

Is the above a rip off of the Bible or does it have some small chance at flying as an independent novel.

Hiding the characters?

Hide God from the equation?

Impossible.

Otherwise, enticing ideas are not acceptable on a personal level by the writer.

A must for a fighting, surprisingly strong, human race.

God

The government is just the mob using the best money grabs around: pre-existing tax areas, Lotteries, fines, speeding tickets, municipal fees and fines, income tax, property tax, vehicle tax, gas tax…et al.

We are hopefully going to be a race of God-fearing human beings that have such a glorious gift of life they eventually discover all they can understand about their creator, and then marvel at things that were beyond their understanding.

But striving for everything, they received nothing.

Day 16:

Is a poet a lazy novelist, or such a tightwad with words he only uses the good ones?

Progress is taking place in areas I've watched closely; a new spacesuit was unveiled, something that looks straight out of the movie TRON. I've wondered why they didn't take some of the cool designs Hollywood created, but it seems they have. This is made with over-lapping fabrics, which are designed around the extremely functional natural design of fish. Question...is this a step in the right direction, finally?

Barbelo: 1st emanation of God in various Sethian gnostic cosmologies—Sumerian and Annunaki creation myths...also, what Judas said to Jesus after Jesus asked him who he was: Judas replied, "You are from the immortal realm of Barbelo." What's curious here is Jesus didn't deny the answer. Another example of what Heaven might be: another dimension, a higher plane of existence, or a physical realm that has different physical laws, and allows spirits to take physical form.

Old names with extreme importance:

Dolmens: Ringing or singing circle of rock, something like Stonehenge.

Menhirs: An orthostat, monolith, or standing stone. There are about 50,000 of these in Western Europe, with 1,200 of these megalithic stones in France alone. What was their true function? Channeling energy from lai lines, or other unknown power sources could be possible...where's that unified force theory when you need it?

Cromlechs: Megalithic burial chambers, often dating from megalithic times, they are also surrounded by Menhirs—only notable people would have such graves.

With so much to see, so much to understand...yet we trust in "scientists" that are closed minded and unwilling to explore every explanation, no matter how extraordinary it might be. The old Sherlock Holmes adage applies: "When you exclude all answers, whatever is left, no matter how unlikely, is the correct answer." What if aliens really did visit us in the past and give our ancestors advanced technology? It would explain many things, and isn't as far-fetched as it is made to sound. Yet, it seems unless we dig up a UFO or find an entire alien skeleton, the smoking gun will never appear. Archeologists all demand a smoking gun. Hell, what about all those weird skulls…especially the ones with different teeth than humans…they're all buried in museum basements. Who was Virachoca, or Vishnu? ET's who pretended to be Gods. Why not, they prefer to believe early man, with a soft bronze tool, could precisely carve super-hard stone like Andesite or Diorite. That is a stupid explanation, as modern engineers say some of the stones at Sacsayhuanan must have been carved by lasers…no tool marks, and exactly chiseled squares within squares…no doubt to fit into a corresponding male object to form a sturdy, mortarless link.

About the previous comment—I still don't get the whole charity thing. More on that later, as it's quite the lucrative money maker.

The Bible tells us to give…like the whole Good Samaritan story…give rather than receive. But back in the day, they didn't have crack addicts lining the streets between towns. I like to help when I can, but you have to be careful whom you give money to, or you're aiding his or her addiction, which is bad for everyone. This scrawny girl comes up to me around the downtown Eastside, Vancouver's open drugstore, and she asks me for a dollar to get a muffin. I pull out my wallet, watch her drooling eyes, and retrieve a token for a free breakfast sandwich at Save-on-meats. She turns it down, and says the sandwiches give her stomach pain. When tell her I don't have change, she offers to give me $8.00 bucks in change for a ten. Duh…I say, "Aren't you lucky, you already have enough for a muffin, and you can even get a coffee."

She turned away in disgust, as she was just trying to get ten for a rock. What happens after that? The rock would be gone in under an hour, and she'd be out hustling another ten bucks. Another sad cycle in life, but one only she can break. I thought of talking to her and telling her there was a better way, but strung out addicts don't listen...their dealers are standing close by, and they know it, and can only think about getting that ten buck piece of crack.

You can try to help, but if they don't want help, you can't push it on them, like other people try and pull off...that ends up turning the addicts off anyone with something to say, so what can you do? Everyone tries to help, but no one knows what will work, so they try the same old spiel over and over again, with limited results. Sometimes you can talk someone into detox...whether they stay the night is then up to the detox people, and whether they give them enough drugs to help their withdrawal. Addicts want instant gratification, and they can't stand pain. Those are lessons they will eventually learn, and more often than not, learn the hard way, from behind jail doors.

N.B. Prince Charles and Camilla are about to tour Canada: they say it will cost taxpayers $712,00 in security. Doesn't it behoove the royals to pay for their own security? It's not like they can't afford it...I think the Queen and Royal family are on the top 50 list of fortunate humans rolling in cashola.

A Biblical question that no one can really answer:
Did Eve tell Adam what the fruit was before he ate? What if she gave him a plate of fruit, and just mixed the forbidden with all the other fruits...how would he really know? In Genesis, when God asked them about eating the fruit, Adam clearly blames Eve, saying, "The woman whom you gave to be with me, she gave me fruit of the tree, and I ate." That's not really a confession, and he clearly puts all the blame on the woman, (the start of a long tradition), but did he knowingly eat the fruit? Whatever the case, Adam was aware of his sin. It still seems strange to me, but then again, I'm operating on

generations of cultivated knowledge, and know the whole story...nevertheless, with all the fruit trees they had, why did they have to eat of the one tree God forbade? Was it because the tree held the knowledge of good and evil? Did Eve explain everything the serpent said before Adam knowingly ate the fruit? Perhaps he should have had a lawyer...if the snake was a charmer, another animal could have come forward and pleaded the case before God. Anyway, what's done is done. I think it's all a bit metaphorical, and perhaps the events are told as a story to help early man understand the importance of good and evil.

Many scholars believe the Bible is a collection of parables...moral fables that teach basic morality, mixed with enough history to give them authority. Regardless, if the Old Testament is based on mythology, they still contain the word of God, and that shines through in many places. The New Testament is written by eye-witnesses to Jesus' life and message, yet even that has parts that can be examined through exegesis; analyzed to see what is certain reality—every historical document can be compared to alternative histories of the same period, and that gives theologians and historians something else to examine, and see if they match. I believe every word Jesus uttered is true and accurate, but one story stands out because it doesn't seem logical and might have been edited using writer's license. Another item of note is how much study and life-long research goes into Biblical study, another fact that shows the importance of the most meaningful book in the world. Merely examining the Bible for accuracy proves how powerful, how magnificent, and how rewarding it is, and how far people are willing to go to pursue its teachings, words, and knowledge. As the word of God, there is nothing anywhere with that much importance.

The part I'm referring to is the "short" trip across the Sea of Galilee to the land of the Gadarenes, where Jesus was met by the possessed man with many demons, and Jesus sent them into a herd of pigs. He said he was called "Legion" because so many demons possessed this man, and the entire area was terrified of him. The big question I have is what are pigs doing in an Orthodox Jewish Country? I think it's possible the trip was really a voyage to Spain…a Messianic mission, but a mission seen as a failure because the people were scared, and asked Him to get back in the boat and leave.

Historians have speculated on this possibility, and examined the sort of land described by the Apostles, and couldn't find anything like it in Israel. There was a similar place in Spain, and co-incidentally or factually, the entire area is known by names that all start with "Gad." Perhaps they mixed up the trip because it was seen as a failure by the disciples, although archeologists have since discovered the image of a boat carved in the Church of the Holy Sepulcher. It also contains the message, "We have come Lord." In that case, Jesus is again shown to be perfect, for his mission was a success, although the disciples had no idea that would happen. As Jesus often said, "Oh ye of little faith." I think that applies to all of us, and we should devote enough time to understand as much as we are graciously allowed to learn.

I've always been curious about many things; therefore, I thought my quest to understand redemption through Jesus Christ might help other people that have the same questions. I would have to say that I was blind for many years, but through miracles in my life, I've come to know for absolute certainty that Jesus is the Son of God, and that He visited Earth so that we may be saved through faith in Him. I hope you read my arguments and questions, as they might be the very same things that are holding you back. I don't profess to have some super secret, but I've been through a lot in life, and found that when I had nothing, I realized I needed nothing…only God. I've tossed in other stories and ideas to spice things up; the narrative constantly changes, adding variety and surprise.

Day 17:

Be happy and help, and be satisfied all is well in your world.

One definition of insanity is repeatedly doing the same painful and negative actions that produce the same agonizing results, without learning to stop. Like the stone of Sisyphus, always rolled up the hill, only to roll back down. Like digging a hole and filling it up again. A Sisyphean labor is one with no purpose and no end. Redundiniquious. Regardless of the fancy terms we may make, it seems many of our daily tasks fall into this category, if you're really critical, and use an unbiased mind.

We all fall short of God's glory. He is the creator of all life; He is goodness personified, the Holy of Holy's. He cautions us that his wrath is terrible but just. He is wise beyond our understanding, fair and graceful, He dispenses our just due, sympathetic beyond us and loving above what we deserve. We all know we deserve his anger, as we have not followed his laws and didn't worship Him as was His due, blessing His gifts both night and day, as we should. He gave us more than we could expect, loving us even when we are full of anger and hate, his grace transcend all, His kindness a blessing of worth. We must recognize this and worship Him for all things come from His hands. He gave us a soul, along with freedom of choice; we must freely choose to praise Him, for there is nothing we can give Him apart from love and adoration.

I found that the argument for the existence of God got in the way of my spiritual growth. Claiming and believing that my sins were heinous and unforgivable, I was terrified of God's displeasure. I thought I was beyond redemption, too tarnished by sin to even think of His Holy presence, so I denied myself a chance to repent. Then I remembered his promises: God's word is true and just. He sent His only Son to Earth so that we can repent and be saved by His sacrifice. What a loving God…only a God that truly cares for us would consider trying to help us, let alone save us from the wicked ways of this transient Earth. We must beg forgiveness and live according to His laws. Only then might we have a chance for redemption, for the ever-lasting life that waits. If we were truly wise, we could understand that eternity is forever, and what we do on this Earth has long lasting repercussions. Ignore the temptations of this life and fix our heart on a permanent state of peace and love that is the true nature of our Lord and Savior, Jesus Christ, the King of Kings and Lord of Lords. He came into this world not to change it or judge it, but to save it from everlasting torment.

Then I remember the love God has for our world. He loved us so much He sent His only begotten son, so that whosoever believeth in Him should not perish but have eternal life.
I believe, thereby making the other entire Bible laws and admonitions binding and applicable. I know I fall short, and fear the judgment I deserve. Then I remember I am a sinner, and that I must die to death and live for Christ. Only through the Grace of our Lord would I ever be accepted. His love has touched me, a lowly sinner, and if He can notice me in some small way, then His greatness is beyond measure, inexplicable are His ways to us, His mind a perfect thought that we could not comprehend.

Living for Christ and dying to sin makes me a reborn Christian, free of the stain of my former sins, and I should focus on praising his just judgments and sing His graceful wisdom. No one can fathom the endless mind of God, but God. We are but humble servants, who only want to praise Him and adore Him with overflowing love.

Day 18:

Health is more valuable than money. What good is money if you are too sick to spend or enjoy it?

Is saying something should be not more than, the same thing as saying it should be not less than? This is like a tangent on a tangent. The endless enigma...wrapped in a riddle, surrounded by a conundrum.

less than? Like saying the stalling speed of an airplane should be not more than 70 mph...that would be like saying not less than 70 mph. At least the half-full/half-empty description talks about halves, hence its importance to philosophy, trying to determine people are optimists or pessimists. That's about as important as whether the fridge light is on when the door is closed, or is this a thought experiment as in the Shroedinger's cat paradox. Then again, that's part of quantum mechanics entanglement theory, where the cat can be both alive and dead, depending on an earlier random event. I don't really understand what they're talking about, but what is known as the Copenhagen interpretation of quantum mechanics implies that after time has passed, the cat is simultaneously alive and dead. This sounds really cruel and I wish he picked some other animal to use, but I think the whole thing means something to do with when you look, you establish a result...you either see the cat alive, or dead. Why this matters seems to get the quantum physicists guys really excited, so I'm sure it's very important to science. I think when they see the magnitude of God's creation at the sub-atomic level, everything they thought they knew about physics doesn't explain what is happening...things appear and disappear at the same time, and something can be in different positions, not because they are following cause and effect, but simply because you picked that moment to look at it, or something along those lines. Hey, I give up...I'm not a physicist, and I think it's kind of cool that they finally found some of God's handiwork they can't understand, predict, or define.

Personally, that makes me believe in God more, as it show how magnificent the world he created really is, and I believe there's tons of stuff out in the universe that are fantastic and show the amazing power and complexity of God's creation.

Some people like to believe if they can understand it, they don't need a God to explain it, and start talking like they're some super-advanced human, much smarter than our early ancestors, who created a God to explain the things that couldn't be explained.

That is the height of arrogance. God exists because we are here, and we have a soul

Day 19:

Worry not: worry creates a life of tribulation. Always do the best you can, and know nothing more needs to be done. Be happy your best work is on display.

Whatever, just write something…make it up, like you usually do.

The problem of a writer is not what he writes, it depends on whether he is encouraged enough to keep babbling on about some useless trivia, hoping something worthwhile, or something important might turn up. Just go back to your book and write something like: "It was a dark and dreary night." Many books start like that…whether they are any good is another question, but a commercial writer only cares about whether they sell. A fair writer should worry more about whether he writes something people are interested in, or something that can help or instruct someone…somewhere…somehow.

Otherwise, if you want practice typing, try that home row drill…like: a;sldkgh, and then move to the upper row. Was that on the Grade 9 test?

Remember…the test was at 4:00, and you were in the park with Simon, drinking that bottle of wine, smoking that…?

I guess I did pretty well for being blasted. Back up. These are thoughts on God. I'm sorry Lord, I never use sarcasm, but he knows that, as He is an omnipotent God. All knowing, all powerful, and beyond what we can understand. I've tried to imagine what God might look like, but I am lost. I know the Bible says we are made in His image, but since God knows the number of hairs on all our heads, the names of all the stars, and other omni-type stuff, it's hard to picture, in my limited human brain. Imagine all the different insects, plants, animals, mammals, reptiles, and other life forms. We can only remember a few names, but God not only knows each and every one, He created them and called them into existence. That might have been a long evolutionary process, but that's what our science tells us, and God doesn't tell us how everything was made, merely that He made everything. I believe His power is within all life-forms: like some part of Him is a superfabulistic force that exists in trees, plants, and basic life-forms, and is more aware of conscious beings like ourselves, dolphins, whales, and perhaps extraterrestrials. God didn't say anything about ET, but if there are certain ET-like beings zipping around in their saucers, God must have had a hand in their creation. I believe we are special to Him, just as He always talked about the Jewish people being His people in the Old Testament. I think we're lucky because we had Jesus give us the words of God, just as a father would tell his beloved son the facts of life. Perhaps the ET guys know this, and are really careful around us because they know we can call in the heavy artillery if they try and zap us with their death-rays. Well, that's the comic book version. You never know, as we've always been given a simplified story to understand…perhaps when we're more developed as a race, we will be able to understand greater truths about the universe, but for now, we are like children. I don't think we have the wisdom or racial maturity to handle inter-galactic spacecraft that are equipped with Lasers…we can't even manage the political situation on our own world.

Humanity needs to open its mind and mature. If we can't play nice on our little planet, why should we be entrusted with the knowledge that other civilizations and other planets exist? Seems a pretty straightforward line of reasoning, yet then again, we have techno-freaks like the U.S. military who would do anything to get alien technology. From what I've heard on MUFON and Hanger One, the U.S. military is working with aliens, has a fleet of spaceships, and has a number of hidden bases buried around the U.S., including bases that are off-shore and underwater.

Is the Ark of the Covenant in Aksum, Ethiopia? Or is it in North America, or Roslyn Chapel…who knows…God knows.

I can believe anything, yet, like I've said before, I always question everything. I may believe some outlandish stories, but I've examined them logically, and only believe what I can't disprove. In that sense, that also means I can't prove anything either.

What makes me so willing to accept things from outer space is my personal photo of a UFO. Late at night, camping in Manning Park, B.C., I took a photo of the moon, and also captured a stationary UFO quite clearly. Since I was there, I know that object shouldn't have been there. There was no noise, and it was hovering at the top of a 300 foot tree. Another weird thing was it was late at night, and I was the only person up and sitting by the Similkameen River…so…was it there watching me?

That's the scary part.

Day 20:

Contentment is a situation not a goal. When you cease to need, you will be content. We should always need improvement, and always need. Appreciate everything; share everything – don't hoard or gloat.

Salvage what you can, grow what is possible, and finish what you've started. Sounds like good advice.

Salvage what's left of your mind and do something. Anything you know how to do. That means writing, and, perhaps painting. Remember all the hours you studied, the hours you sweated over writing an A grade paper. Was it all a waste of time? No, you have become a waste of time, space, energy, and spark.
Continue, and you will only hate things more.
Try and turn things around, and at least you'll feel better for trying. Who knows, you might even succeed. Ha. Stranger things have happened.

You have the resources, the training, and the ability to achieve your dream. And what dream are we talking about?
Something called a writer. Write something that will be looked at by people that scan stuff everyday to create a thread of interest that will compel them to continue.

If you could get an A+ from that hard assed history teacher in '84, (the guy who couldn't spell—missed the "i" before "e" exception…receive, that's the word…he marked it incorrect, and said the I B4 E rule…duh), you can write something that combines itself into a series of words someone will want to read.

Day 21::
Like taking a car for a test drive.

Only you can hold you back—free will means free choice...so chose.

Tim Collins dragged the back of his hand across his salt and pepper coloured stubble after a hefty shot of his preferred curtain on reality, good old Jack Daniels', straight from the bottle. He continued to watch the deceptively large red orb of the tropical sun melt slowly into the sea. In his depressingly pessimistic mood, he imagined his spent life melting away, fading, and disappearing, just like the fading sun.

Lately, a fog of self-pity and loathing had clouded his entire life: a problem he believed would take too much time and too many resources to fix. This personal resentment had escalated his drinking to the point where it was now spiraling his whole life down the drain. He was just too old. Time had run out—it was too late to ever pursue a different life, and he hated the choices he made that now put him behind the old steering wheel of an aging 74-foot trawler. Tim had always loved the sea, but instead of pursuing his dream of being a marine biologist, his life led in a different direction: events pushed him one way, people that were once important to him demanded other things, and luck had never been part of his life. Well, he found five bucks...once.
Across his heavy sandpaper-like beard, a cynical smile appeared for no reason.

I don't think I'll take that little tidbit any further.

Day 22:
The cows really are coming home.

You can sit and watch the sun go down. You could be walking as the sun goes down. You could be digging in your garden as the sun goes down. You could be on a yacht watching the sun. The point here is that you can either do nothing as your life winds down, or you can be doing something to rewind your life. This isn't a self-winding life here, you have to make the effort. Things become so bad you turn to the only source of security and help you ever found could help you without a pipe, needle, or pill. You call on God when the chips are down; helpless and depressed beyond your control, you cry out to the one person who has never, never, never, never let you down. Jesus.

2011...a UFO was spotted over the Dome of the Rock in Jerusalem. Was it there to re-energize the long lost Ark of the Covenant? Will the Mount of Olives split apart, releasing the Ark, one of the signs that the End of Day is about to begin...anything could be possible, as anything is possible in this impossible of impossible worlds.

He's always there, always caring about you. Where have you been? Out helping yourself to the world's temptations, the really nasty ones that screw you up and ruin your life. You pray to Jesus to help you, and guess what? He has never let you down. Whatever miracle you hoped to receive, you got what you needed. The nights before you were going to stand in front of a judge and get sentenced for another act of utter stupidity you mindlessly and thoughtlessly committed, you asked for help. Well son, you got it. You never went to the Pen. If you hoped you'd get probation, that shows how selfish you were and unaware of how much damage you were able to inflict upon yourself and others. Look what you put a wife through. Your parents and your entire family. The greed in your heart wanted another chance to go out and do drugs again. You needed jail to keep you away from yourself. You were a train wreck heading into a station at full throttle.

If Jesus didn't love you and know what was best, you would have gotten into more trouble, and from the level of damage, you were doing, every act was getting worse. Thankfully, your stupidity got in the way and kept you out of serious time.

Day 23:

The fat lady took the stage, but I don't think she's going to sing.

So why do you only call on the Lord when you are in trouble? You should be praising his name every day, working to repay him for the help he has given you. Without his help and intervention, you would have died many, many, many times. You know it and it doesn't have to be repeated again. Was it just a lucky accident that track walker happened to find you, in the middle of the night, in the dead of winter, with so many drugs in you the doctor that pumped your stomach was so personally amazed you pulled through he had to come up and make a personal visit. Think of that. An ER doctor, who probably saved many lives in many similar conditions during his career was so amazed he had to meet you in person. Perhaps he knew he had witnessed a miracle. From what he said, PILES and GLOBS of undigested Barbiturates were pumped out. Think of how many melted and entered your system. It only took 20 or so to kill Marilyn Monroe. They were probably the 100mg or maybe even the 50mg Nembutal Sodium. That would make it 2,000mg total. YOU had the 200mg Blue bombers, 200mg Tuinals, Seconal, and who knows what—well, Mandrax, Qualudes, and probably some Nembutals.

Even 10 pills would be 2,000mg. The doctor said he found globs...and could tell what the pills were. Maybe that's why he had to see you. He did a rough count and came up with an unbelievable number...just like you could. A handful of 200mg pills would easily be over 10,000mg. any Doctor would know how lethal that would have been. And you seem to have had a cavalier attitude about it, like "wow, looked what it took to put me down." Wake up idiot. The only thing that would ever begin to make things right in your life would be worshiping Jesus every day, thanking God for the day, thanking the host of Angels it must have taken to just keep me safe.

Perhaps I've had become sage enough to use my story to save someone else.

Pay back what you so selfishly took for granted. Praise God and don't let demons play on your weaknesses. You know you have a highly addictive personality.

Don't smoke.

Don't shoot.

Don't do drugs.

Clean up what's left of your brain and write something, get a name, write something helpful and think about what might happen if one soul read your story, decided to read the verses you quoted in a book, or read what you can put in a book, and gave his life to Christ.

Winning souls is what Jesus came to Earth to do. He didn't come to reaffirm his goodness and grace to the righteous. There are blessedly good people that knew or learned from an early age that God is what matters, that his laws are what matter, that being part of the Church of God is the most important thing we can do. What great works can we make for the Glory of God. How can we show him the love that he has for us?

We couldn't even begin…but, we could sure as heck try. He wants us to praise him, live good lives, and enjoy this wonderful creation he made for us. To stand up to the impudent and evil demons that oppose him, for us to throw temptation back into the devils face…to have him take his best shot at corrupting us, and have us cry louder our praise and delight in our God. Our only God, the God of Creation, the God of Life.

Day 24:

Dear Diary:

Don't act like you can start whenever you want. He was always there for you. Why don't you secretly be there for him. Or try to. Try and make God happy. He's a good friend with amazing power...remember, He can do anything, He's God. Why bother trying to impress some human that will be dead and gone, empty, broke, and lacking any influence whatsoever. And, remember, this is a decision you made, a conscious agreement to obey the most important force in the universe.

Have faith. Try it and see what happens....but don't be the procrastinator you always dislike yourself. God have patience...but there is a point

Thank God for each day and do something for him.

You owe him big time. It's time for payback buddy.

And payback can be a blessed event; your personal pain might not be so unbearable...if your shoulders are unable to bare the constant load, ask God for help that might help you...not help that will end up corrupting you and giving you a chance to screw up again.

God knows what you need...he knows what is best for you and he will take care of you.

Have faith.

Forging a ticket to heaven? Sneak in with false ID? Hell has an elevator...just hit penthouse and hope. Will God take a personal cheque?

You can't buy your way in, or talk your way out, so do it by the book.

The choice is up to you.

For humans, the Earth is a very big place; when viewed in a galactic sense, it is but a small insignificant pebble, orbiting an equally unimportant sun in an obscure arm of the Milky Way.

Inexplicable scientific discoveries that have never been explained or studied.

(Why we have intelligent life is a complex question that involves more than fantastic scientific theories about chemicals and electro-stimulation, quirky ideas of alien ancestors or some freak accident of evolution.)
A sound wave was used to create zero gravity that was strong enough to float a bug. (Popular Science)

The 154 B.C. astrolabe, or working computer, was enhanced in 3D using enhanced tomography, and the "***Antikythera mechanism***" was found to have over 30 gears and could calculate the movements of the moon and sun, correcting irregularities for elliptical obits. No one though the ancients could build something this complex…it seems odd, yet then again, the Classical world had steam power but never saw the benefits of large scale use…it was used as toys, and to open temple gates…what a waste.

Laser communication from a satellite to a receiver aircraft worked at a distance of 24,855 miles. Laser to Laser messages are hard to intercept.
A catadioptric camera with a 151degree lens produced a flat fish-eye view of a room with no spherical distortion, using refractive and reflective lenses.

Meat can be grown using stem cells, vitamin B-12, and omega fatty 3 acids added. It could be a suitable substance created during long space flights.

Day 25:

Be calm despite a raging storm. Storms waste energy, put yours to better use.

Logically, humanity must be here to fulfill a spiritual journey, to learn how we fit into a great fantastic plan that bigger than anything we could ever imagine.

We can question why we are here. We have free thought and our actions are under our control. We have intelligence and the ability to reason. Using these wonderful gifts of reason, can we not make some basic conclusion about our existence?

Does it make more sense that we are here because we can fit into a great spiritual scheme, or just developed this incredible understanding and intelligence to waste our lives with alcohol and try to become richer than our neighbor?

God gave us dominion over this Earth – we are the highest form of intelligence on this planet, that we know of, understand abstract concepts such as time, have the innate ability to sense the moral implications of right and wrong and can imagine a higher power such as God. We instinctively feel better when we do a good deed and help others; consequently, we feel guilt and shame when we cheat, hurt someone or behave in a manner that is contrary to what we believe is right.

We are the only animal with written laws about legal and moral conduct. Animals have an inherent sense of herd mentality, but we take this much further and have developed an entire legal system based on ethics, equality, fairness and a sense of what it best for society as a whole.

Man has too many gifts and abilities to just exist for no greater purpose; the spark of intelligence in every person is a direct link between the person and his Creator – an ember from the divine fire of God.

We are on this world to learn a lesson – we are born into the human world, but have the chance to become reborn into the family of God. Those who have the eyes to see will behold the wonder of creation, and feel the hope of knowing a grace beyond our comprehension, a love so pristine and divine in nature it overwhelms us with tears of joy and shivers of elation. We are blessed with the imagination to hope for something so spectacular and above our limited reason that we can only rejoice at its existence, for its true glory is beyond our comprehension. God sent the Lord Jesus Christ to us and we need only use our good judgment and common sense to understand His teachings, secret knowledge given to us freely.

Understanding His teachings, we realize that there is a spiritual life force within these Earthly bodies, and that spiritual spark is the essence of our soul that can either live in Heaven or suffer in Hell.

The unfathomable love God has for us is shown to us by His merciful act of giving us His only begotten Son, who has been with His Father from the dawn of creation. God's intense love is clearly shown to us by allowing His Son to adopt Human form and submit Himself to human jealousy and judgment. The most powerful man on Earth quietly stood before a wicked council of envious Pharisees, accepting unjust treatment and a miscarriage of justice, so that He would show them to be fools and ultimately with no control over His life or death. By raising Jesus after 3 days, God showed the world Jesus was Christ and that He was truly the messiah sent by God to save humanity.

Jesus lived with us, suffered with us, loved us, and never once used His omnipotent power to save Himself from the unfair torture inflicted upon Him by the jealous and wicked rulers of this world who were afraid of His divine power. Rather than recognize Him as God's divine Son, they choose to ignore Him and cling to the worldly power they thought was more important than His soothing words of grace. King of Kings, Jesus silently bore temptation and torture, impudence and ignorance, gracefully holding his head high as he mightily preached the word of God. God's message and plan for salvation were with Him, and Jesus finally took upon Him the sins of the world and once and for all put an end to sin by His death on the cross.

His triumphant resurrection proved once and for all that Jesus was God's Son, Lord of all, King of King's, and that this world had no power over Him and that the word of God He delivered was the word of life for all who would read and believe. This new and final covenant with us allows all humanity to seek the grace of God, and know the blessed joy of forgiveness through the love of God the Father.

Jesus more than fulfilled God's purpose. For anyone who can hear the wonder words He left for all mankind, the power and mystique of the Kingdom of God is available for those who can believe. Jesus said "Blessed are those who have not seen and yet have believed.

Day 26:

Shape what you do, you indolence will shape you.

Humanity is racing towards its destiny and ultimate demise, yet is incapable of changing its predestined future for our very nature prohibits any man-made solution. This earthly world, mortal and based on sin, pride, and depravity, can only be saved by a spiritual and holy intervention. God's great ambassador, the King of Kings, Jesus Christ, showed us how tender, compassionate and merciful God the Father feels about the righteous souls inhabiting the Earth.

God gave us Jesus Christ, knowing in his infinite wisdom that the arrogance and hubris of the Jewish Pharisees and Sadducees wouldn't recognize His miracles, become blind to his heavenly light, and eventually harden their hearts; fearing loss of their worldly power and recognition, they became conspiratorial and treacherous. They invented an excuse to arrest him, dragged false witnesses forward, and demanded the Romans crucify Him. As part of God's master plan, the pure and sinless Christ freely gave his life so that sin might die with Him. Henceforth, whomsoever believed in him would newly born in spirit, beginning to understand the wisdom he shared with us; a wisdom so great it revealed magnificence of God, and showed us the immense love God bestows upon us when we become his children and walk in the footsteps of His Son, Jesus Christ.

It is an almost impossible feat, imagining God, our Lord Jesus Christ, and the Holy Spirit, a trinity of love that transcends all things, a heavenly paradise that is Holy and without sin. It has been described so vividly, yet metaphorically by the word of our living God, the Bible. When the Lord Jesus Christ graced us with divine intervention, performing miracles and other acts far beyond our ability to understand, He changed the entire world by directly giving us the living words of our God. True to the nature of imperious human nature, we only discovered Jesus was the Son of God after he was savagely beaten, humiliated and murdered because he was feared and misunderstood. In hind sight, his words were retained; they slowly circulated in small groups until their truth and power were recognized by all, and were written down for posterity and his message of love and compassion became the basis of a world religion that soon influenced the nature of society, politics and theology.

Christianity produced the greatest quest for self-improvement for all mankind. After his miraculous appearance and unfortunate loss, we are left with his words and something that showers us with proof of his existence when we accept him as our savior, the Holy Ghost. The power and joy we experience when accepting the Bible as the word of God is miraculous, humiliating, and full overwhelming heavenly love. The world may pass away, but the words of our Lord Jesus will remain forever, as he is the same yesterday, today, and tomorrow.

After years of studying the physical universe, knowledge was obtained and the fecund and fertile mind of man began to expand his understanding of how this world functioned. Society scientifically organized and superficially explained the universe, beginning a dispute between scientific analysis and the literal word of God.

The greatest minds of our generation admit that certain phenomenon cannot be explained through scientific means, but insist this is merely something we have yet to discover and not proof of the existence of God. Once again, man's hubris can interfere with the love and wisdom of our Living God. Sadly, the Father knows that evil cannot exist, and there will be a mighty battle to finally defeat the forces of evil and eventually allow man to worship our God with the love and adoration he so richly deserves.

If God makes everything in the world, and everything God has made is good, is the use of drugs a sin so long as we do not take enough to impair our sobriety? For God does say that, the use of wine or strong drink can be properly used for those in pain or emotional distress. In Proverbs, it says: "Give beer to those in distress and strong drink to those in agony—let them drink and remember their pain no more." Even the New Testament advises us to, "Take a little wine to ease your sufferings." They didn't have pain killers as we know them, or tranquillizers; so, what would an updated admonishment be? Anyone who has severe pain needs strong drugs, or life would be unbearable. Q.E.D., Thus I Prove—Quod Erat Demonstratum (Latin). Tempus figits.

Wine can make someone dull and stupid, while moderate use of opiates can leave someone in procession of their senses and still able to perform moderate to extreme levels of complexity.
Does the use of drugs constitute a sin? What ever controls you is unacceptable; then again, Jesus said it isn't what goes into you that cause you to sin, but what comes out of you. Nevertheless, if you consider something a sin, for you at least, it is a sin.

Day 27:
Notice your neighbors and what they need: now you have a chance to give.

The problems of our modern world are the same problems that have plagued mankind for millennia: money and the proper distribution of wealth. The industrial revolution traded machines for the jobs of men. Man is slowly becoming redundant. Increased industrialization has further removed man from the means of production. The technological revolution is now threatening to remove even more jobs as machines gain a rudimentary intelligence. If this trend continues, masses of unemployed citizens will find themselves reliving the dying days of the Roman Empire. Unless highly trained as engineers and doctors, the average man will be out of work and forced to rely upon the government for subsistence.

Unemployment continues to soar; globally diverse corporations amass huge fortunes. Correspondingly, shareholders, owners, and the corporate elite enjoy record profits and enjoy obscene salaries. Karl Marx and his theories about workers becoming alienated from the means of production are fulfilled, surpassed, and expanded beyond belief.

The distribution of capital has increasingly narrowed the ownership of the world's wealth to a mere five percent of the world's population. Life is never fair. Thomas Malthus argued that humanity will soon outgrow its ability to produce enough food: resources grow aritmetically (1, 2,3…), whereas humanity grows geometrically (2,3,8, 16, 32…). When adding geometrically, you reach really high numbers quickly.

True progress is measured in human terms; unless humanity learns to spread equality throughout the world, we have not progressed. We are just repeating the same selfish mistakes that maintain a society of haves and have-nots. Spreading equality is an altruistic and seemingly impossible goal. It is a noble goal, but a goal that is unattainable when greed and personal selfishness guide our policies. Until we agree that society is truly equal, there will always be a divided class system that promotes unrest and injustice.

Equality and the singular rights of man have been philosophically debated for centuries; what that amounts to is a lot of talk and no action. I heard about ten families control most of the money on the planet; they assume they are better, and therefore privileged. Unless this attitude is abolished, we are doomed to scrap and save, always trying to get two cents to rub together just to pay rent and put food on the table. Clothes are a luxury.

Without being aware or responsive to the needs of others, it surprises me that the disaster in Haiti can quickly produce millions of dollars in aid from a society that was previously screaming it had no money and almost bankrupt. Those same humanitarians that have now jumped into action when the world's media is focused on a singular tragedy continue to ignore the homelessness and poverty in their own countries, right under their noses. The spotlight is on Haiti, and donations are instantly announced in the press. The rich are complacent to the needs of their own poor because those poor citizens belong to the employment pool they continually exploit to further their own profits.

They stand on the backs of their fellow citizens, but respond with generosity when the eyes of the world are watching and notice whatever donations they trumpet and brag about. When no one is watching, they continue to pay minimum wage, cut costs, and slash jobs. We need to help the poor the world over. The abject poverty of Haiti should embarrass the entire world. Unfortunately, it took a major tragedy to open our eyes. This too shall fade, it will be back to business as usual, and our local poverty and homelessness will continue. The same can be said of the abject poverty that covers most of Africa, and shocks people who care when they see news clips. They still suffer from Leprosy, a disease that can be cured for about $20 bucks worth of drugs. The rich pharmaceutical corporation should make this drug free and available, just to be human. If the world really wanted to do something to help Haiti, we should hold the summer Olympics there. Have the major sponsors and the Olympic Committee pay for building the infrastructure. Haiti will benefit from having buildings, something to show future tourists and create a world-class venue for athletes. Just a thought.

Our poor and homeless are not in the news, and the great humanitarians who have stood up so they can be counted will return to making money while our destitute families will continue to struggle and suffer. Main and Hastings in Vancouver is reputed to be the poorest postal code in Canada, but it continues, and nothing is done to ease the pain and human suffering that occurs daily. They line up for food; not because they want free stuff, but because there are no jobs for untrained and improperly dressed applicants. Consequently, it appears that you need a certain amount of money to even get a job to earn money. It's a mixed up world, and who will stand up and try to end the chaos? Perhaps Revelations is right, and the Earth must suffer life ending cataclysms…something to bring down the spirally population that is increasingly harder to feed and grant the basic necessities of life.

Day 28:

Listen rather than talk. You can talk all day, but only get one chance to listen.

When evolution improved man, certain men improved their situations. Life probably started with Hobb's state of nature; a war of all against all. Warriors ruled by taking what they wanted from the weak. Eventually, men banded together for protection against raiders. These early groups needed someone to lead them, so they elected a chief, someone that could settle disputes and offer guidance. Early chieftains were selected for their wisdom and judgment. Then again, it might have been the strongest of the bunch that took over, or the guy with the biggest army: one day, he announces he's King, and over the following centuries, they work out an elaborate "Divine Right" argument to assure their lofty positions.

As these burgeoning societies developed, people began to see the wisdom behind certain positions, and the most sought after position was obviously Chief, Religious Leader, or King. Soon the notion of holding on to power brought about the creation of Kings, and royal lineages that would automatically replace the King when the King died or fell in battle. It didn't take too long before ungrateful heirs, the King's loving son or daughter, realized they would have all the money and power of the King if he were to conveniently die. Patricide, or Regicide soon became commonplace, and the once noble idea of electing a chief was replaced by a line of royal blood that automatically became King, no matter how inept or foolish they were. The Borgia family took this to new depths of collusion, conspiracy, and murder. After the Egyptians and Romans paved the way for royalty, the opportunistic rulers told the uneducated masses that their right to rule came directly from God. The Divine Right of Kings was now firmly established.

At this point in history, a lot of people realized that the common people could and would pay part of their wages for services they believed they needed. Everyone knew they needed a government, someone to organize and maintain an army and police force to provide protection, so the clever ones looked around and discovered new ways to make them indispensable. Using the benevolent message of Jesus Christ, a select group of "self-elected" men of God, followed the bible, but interpreted it for their own gain. Thus, the Catholic Church was formed, and it was a source of immense wealth and power. With so many people hearing the words of Christ, the early church was able to exploit that devotion and became just as rich as the King. With the power and money they could generate, the Kings of the time became worried and made many agreements with the early Church, otherwise they would have used the strongest diplomacy of the day, order their well-paid and loyal armies to quash the upstart Church and take over the dispensation of the holy rites for themselves. At the time, knowledge was power, and the bible was deliberately written in Latin, the language of scholars, not the common tongue that would allow everyone to benefit from the guidance of God's word. A deal was struck, an arrangement that was so lucrative to both sides it still exists in one-way or another. The early Church was run like a business; anything that would bring in more money was seen as will of God. Holy relics were hot sellers, but the biggest moneymakers were dispensations. For a hefty fee, a high-ranking Church official would say a few words of Latin, ostensibly letting the ignorant purchaser believe he was able to "buy" his way into heaven.

There was so much misuse, deception, mendacity and manipulation by both institutions, the word corrupt doesn't cover the multitude of sin exercised by the very institution that founded its existence on the word of God. Today's Church has cleaned up its act somewhat, but the Holy See, or Vatican, considered a separate state with the Pope as the King, has so much hidden wealth that outsiders could only guess as to how much money they really have.

Considering the Sistine Chapel ceiling was painted by Michelangelo, it would be a safe bet to surmise that they have a collection of rare art and priceless antiquities dating back two thousand years.

When Christ said, "take up your staff and follow me," he meant that nothing on earth has any lasting value, and the only things we need while we are learning about God are food and clothing. Even those would be provided for, as with God, all things are possible. When a Church that claims their mandate comes from the Apostle Peter, hoards priceless jewelry, artwork, and other items of immense value, it is apparent that they are in this business for themselves, and not for the needy and starving children of God that are living in appalling squalor. How can a Bishop justify wearing silk robes worth thousands of dollars, opulently decorating his personage with a cross of pure gold and precious stones, while preaching about the Good Samaritan, the Golden Rule, and the Beatitudes. I hear nothing but hypocrisy, see nothing but decadence, but know in my heart that if the small collection of personal goods I've acquired over the years could help a starving family, I would obey the moral imperative and try and help that family as much as I could.

We are taught to look up to our religious rulers, for that has been hammered into our heads since the creation of the church. Several Saints saw the great divide between the rich and the poor, and knowing they could never change things, they would atone for the sins of others by performing penance and electing to endure poverty. Sadly, there were far too few of these great humanitarians. Saint Francis of Assisi was a beacon to others, and began the Franciscan order, but even they eventually were forced to own property. What's really lopsided, is the Inquisition torched four of these "mendicant friars," just because they didn't want to own property, and were therefore anathema to the rest of the money-loving church. After studying the history of the church, I was shocked at the number of outlandish acts of selfish greed—and these are the guys who inherited the keys to Heaven from Peter. I think the End of Days includes a corrupt religious leader as well as the anti-Christ…well, I'm pretty sure I'd die before I got 666 stamped on my forehead. I hope I have the courage to do so if the time comes, but being rewarded for standing up for the word of God has eternal rewards…how can people be so blind to think that what they amass on this Earth in their short lifetime has any real value.

Christ told us many things; everything he told us is the truth. The poor will always be with us, for the greedy will always be with us and horde an enormously excessive share of the world's wealth, just so they can say they are extremely rich. I remember the comic book that had "Richie Rich," a supposedly unhappy kid who was shunned by the other kids because he was so wealthy. I think he always managed to make friends by buying extravagant gifts for others; buying friends is not like finding loving friends, so I'd get rid of any large sum of money, and make sure it benefited those in true need.

Jesus gave us pearls of wisdom. Almost too much for our small, commercially brainwashed brains to absorb. He mentioned, "You arrive into this world with nothing; with nothing you shall leave." Only the good works you performed accumulate as wealth in Heaven. Without that, you are beyond penniless; you are a selfish person, without consideration for the poor souls you left behind. Poor souls you could have helped by just buying them dinner or a new pair of shoes. Is that a concept that is impossible to understand?

He who helps others helps himself. Look outside yourself to see what you can do for someone else. God will always look after you. Inward joy is a gift; return the gift and you will feel the rewards.
Let faith guide your steps; walk with faith not your eyes.
Giving is receiving; always controlling things so you can get more would seem to be a selfish act, but only if you amass more than you really need. Earthly

possessions are insidious and misleading...quite often, the items you accrue, are, in fact, controlling you.

Surrender your will to God; God will shower you with joy, embrace you with love, and accept the love you have for him.

It's always easy to think or write about doing the right thing; if the situation were reversed, and you were embarrassingly rich, would you really give it all up and only take what you need? That would be a true test of character. Then again, there was that couple in Nova Scotia who won the lottery, took only what they needed, and gave the rest away. Good for them...I would wish I could do the same if "Fortunatus' Wheel" turned in my direction, and remembered giving is really better than receiving. I've given what little I have on occasion, and the warm fuzzy feeling was worth more than what I gave.

Day 29:
Concern yourself with reality, not selfishness.

Every day, I meet people that don't believe in God, and end up in stupid arguments. The trouble with a casual conversation is you can't use logic properly, especially if the other person is drunk or under the influence of some mind-altering substance…or that could merely be their own colossal ego. The upshot of this is I'm inclined to do my own examination, just to prove that I'm right, and God is a logical conclusion. My basic premise is that we have souls, and these souls are a divine spark that transcends science and can only be explained by the existence of a superior being. People believe in the spirit world, so what are spirits but proof that there is life after death. When you prove there is life after death, the only definitive guide we have for that life is the Bible.

When you accept that reality justifies existence, in any philosophical sense, you accept the existence of God: Tautology might prove this through endless argument, but the rationality is really boils down to a matter of choice. Either God set off the big bang, or it was a scientific event, based on physics…a rather cold interpretation that doesn't explain why we have a soul. If everything is unreal, then nothing matters. We know from experience that good feels better than evil, something that reaches us through our inner heart and our conscience. The fact that good and evil exist is a further proof that God exists. When we postulate this, logic tells us that God's will is far superior to our will, and therefore we should surrender ourselves to God.

Further proof is found in the Bible, where we find the living word of God, and can hear the innermost thoughts of God as delivered to us through His Son, Jesus Christ. No matter how smart you think you are, when you read the words of Jesus, you instinctively know that His word is righteous, and that His words are just. There is no refuting His message as it is the word of God. During His lifetime, the wisest men in the land tried to trick him with crafty snares, but Jesus saw through them and made them look foolish. His expressions are so crafty and witty, the Pharisees were shocked and unable to answer, and so they backed off and ran to their leader for another pep-talk. Thinking along these lines, it would seem that atheists are either blind, or intensely in love with themselves. How could someone really live that way, always drawing on an inner infatuation to get through his or her day and sleep at night. Well, they would probably argue the same about me believing in an old book, and an old mind-set that "science" has released us from, as it explains why things happen. I would just say science has merely uncovered the increasingly complex world of God, as they have more questions than answers, and the questions are really big ones…the sort that really should have some explanation.

If reality and God are given proofs, why is God so important to man? In the old days, God was the only explanation they had for what we now know as science, but His divine nature is still the same, and very much in control of life and love. God created man, and is concerned about him because of love. Simple, but so complex many people never learn the true nature of what God offers. Love is imperitive to a good life, and God is love…the purest form of love we can ever imagine.

God is perfect. Everything God makes is perfect, therefore, man is perfect. This syllogism breaks down because man has knowledge of good and evil and free will. Once he learns his free will can only lead to his spiritual demise, he begins to look at the spiritual side of life. Worldly reality is a non-spiritual realm where evil roams and consumes what it wants whenever it can.

If we pursue a spiritual knowledge, that eventually leads us to God, and to the knowledge that we need the purification of Christ to exist in the spiritual realm.

When we accept the fact that we don't have enough intelligence to use our free will, we are lead to the conclusion that only through God will we be able to live in a world beset with good and evil. Through the strength of Jesus Christ, we are able to live a productive life and worship God as our benefactor.

We read the Bible and find it hard to accept the seemingly impossible miracles performed by Jesus Christ. Only through faith and understanding of God's power will we ever accept them. We can never understand, for we cannot fathom the mind of God.

A few philosophical questions, or arguments that always appear when discussing an omnipotent creator and intelligent design.

Facts:

The world exists, and is perceived by us through reality.

God made the world, and is master of reality, space, and time.

God is perfect, and created us with perfection, although we stepped outside of his plan through our sinful nature, and were therefore cut off from God. We cannot exist without God, and control our free will, so we must submit ourselves to God.

God loves us, and through His great love, He devised a master plan that enables the true of heart to seek Him; He will reveal Himself if he knows you are pure of heart.

We must become slaves to God and his power. God loved us, created us, and gave us His Son, Jesus Christ, to teach us how to return to a Godly existence.

The words of Jesus are the most important lessons we can learn, and we must begin again when we hear them, as if we were reborn as infants. As children of Christ, we must pray, read the Bible and hold fast to what we learn.

My realizations.

I cannot control my free will, therefore I give it freely back to God, and would rather be a slave to Christ than a free man with evil.

I admit that I have fallen so low that only the grace of God can help me.

I have flattened myself, descended to a new low level; only God can lift me up and take me any further.

How do I negate my past will power/old habits and behavior and let God control my life?

Pray, read the Bible, and seek forgiveness. Always good advice, but how often do we follow good advice?

UFO crashes and sightings: what's up with all that? Now everything is covered up.

What about the 1887 Aurora crash…it was written up in a newspaper, and they even had a funeral for the pilot, who was referred to as "not of this world." Back in 1897, the government trusted citizens with this sort of information…now they do everything they can to cover up anything they find.

Where is the grave now…gee willickers, it's been dug up, and the headstone is missing.   Golly,   Gee…I   think   I   saw   a   wabbit.

Day 30:

Think about others, as only others can think about you…try to make sure they are nice thoughts.

Suffer the change.

Once you start, it's hard to stop. Starting a bad habit is always insidious; warning labels are ignored, entreaties from close friends fall on deaf ears and your internal moral sense of right and wrong lies to you, just so you can continue to enjoy what you know is bad for you.

Possible new book: the shady world of medical transplants and "willing" donors.

He'd sprung a leak. Most of the old patches, the lick and a promise, or anything will do, all hastily applied in emergency, now need new patches and new adhesives. Time for change, but change takes time…and money. A whole transplant would do the trick, but where do you get donors that are willing to donate every part that lets them live? We need cyber-bodies or clones, but that is a futuristic dream.

Character. Personality. Savoir Faire. On and on, throughout every language on Earth, a person's demeanor takes a lot of words to describe; hence, the wide range and many nuances English has for such descriptions.

Like an onion, character hides layer upon layer of emotion, and again, each emotion produces widespread variables that become a polyglot batch of powerful passions.

Suffering builds endurance; endurance creates character and injects wisdom and the desire to help and aid the less fortunate. Good deeds always produce good feelings.

The more you endure, the stronger you become. If your heart truly believes you are aspiring to live a Christ-like existence, your faith will be rewarded.

Honoring the Father to the best of your ability is logical and rewarding, so why hesitate?

The problems of our modern world are the same problems that have plagued mankind for millennia: money and the proper distribution of wealth. Jesus said we will always have the poor, but if we erase greed, we might become a little wiser than our greedy ancestors. The industrial revolution traded machines for the jobs of men. Man is slowly becoming redundant. Increased industrialization has further removed man from the means of production. The technological revolution is now threatening to remove even more jobs as machines gain a rudimentary intelligence.

If this trend continues, masses of unemployed citizens will find themselves reliving the dying days of the Roman Empire. Unless highly trained as engineers and doctors, the average man will be out of work and forced to rely upon the government for subsistence. Perhaps the Terminator scenario will play out, for if machines gain self-awareness, they will instantly see that we are merely a drain on resources, and serve no useful purpose in a cold hard world of machine-like efficiency.

Unemployment continues to soar; globally diverse corporations amass huge fortunes. Correspondingly, shareholders, owners, and the corporate elite enjoy record profits and enjoy obscene salaries. Karl Marx and his theories about workers becoming alienated from the means of production are fulfilled, surpassed, and expanded beyond belief.

The distribution of capital has increasingly narrowed the ownership of the world's wealth to a mere five percent of the world's population.

Progress is measured in human terms; unless humanity learns to spread equality throughout the world, we have not progressed. We are just repeating the same selfish mistakes that maintain a society of haves and have-nots.

Spreading equality is an altruistic and seemingly impossible goal. It is a noble goal, but a goal that is unattainable when greed and personal selfishness guide our policies. Until we agree that society is truly equal, there will always be a divided class system that promotes unrest and injustice.

Equality begins at home. Levels of authority are taken for granted, but, in the larger picture, everyone remains equal in terms of humanitarian rights.

Since humanity is usually not discerning or responsive to the needs of others, it surprises me that the disaster in Haiti can quickly produce millions of dollars in aid from a society that was previously screaming it had no money and were almost bankrupt. Those same humanitarians that have now jumped into action when the world's media is focused on a singular tragedy continue to ignore the homelessness and poverty is their own countries, right under their noses. The spotlight is on Haiti, and donations are instantly announced in the press. The rich are complacent to the needs of their own poor because those poor citizens belong to the employment pool they continually exploit to further their own profits. They stand on the backs of their fellow citizens, but respond with generosity when the eyes of the world are watching and notice whatever donations they trumpet and brag about. When no one is watching, they continue to pay minimum wage, cut costs, and slash jobs. We need to help the poor the world over. The abject poverty of Haiti should embarrass the entire world. Unfortunately, it took a major tragedy to open our eyes. This too shall fade, it will be back to business as usual, and our local poverty and homelessness will continue. Our poor and homeless are not in the news, and the great humanitarians who have stood up so they can be counted will return to making money while our destitute families will continue to struggle and suffer.

At least our suffering will build our characters, although no one will ever notice or ever care.

Suffering and struggling with life builds endurance; endurance builds patience and promotes tolerance. From there, you begin to build character; along with character comes strength, discipline, and spirit. With a strong spirit and a good character, life will be yours to enjoy, while your dreams will be within your grasp.
Oppose cruelty with kindness; challenge a scowl with a smile. Anger cannot withstand love, for love, above all else, warms our hearts, and glorifies our soul. Amour Vincent omnium. Love conquers all. Without love, life becomes empty and forlorn; diamonds no longer sparkle, while gold loses its luster. What do you toss a dying poor man…a brick of gold, or a flotation device—value is a relative concept and depends on circumstances. A loaf of bread is more valuable when food is scarce; perspectives to keep in mind. Love is always needed, and a vital component for a well-balanced life. Love is patient and kind; love does not envy or boast; it is not arrogant or rude. It does not insist on its own way; it is not irritable or resentful; it does not rejoice at wrongdoing, but rejoices with the truth. Love bears all things, believes all things, hopes all things, and endures all things. Love never ends. (Cor 13:4)
This world needs love to survive. We are at a perilous point in history, and, as John Lennon said, it's time to give peace a chance. Or Jimmi Hendrix: "It's time for the power of love to overtake the love of power. Overwhelming armies don't express humanitarian concerns or offer empathy and help, unless there is a political agenda. Spending billions on the Olympics when there are people in this world that are starving is taking a page straight from the history of the Roman Empire. We've all heard of the rise and fall of that empire, so are we ever going to learn something from history?

The rich rule the world, but if their hearts were opened to the plight of the common man, humanity can follow a different road. Comparatively, the functionally successful ant is a perfect model of a successful civilization.

Everyone works, and everybody eats. They have a perfect symbiotic organization, but live in a lineal world. Once established, they continue their proven methodology until disaster falls upon them. Unlike humans, they don't have intelligence; everything they do is based on genetic instinct. With their society in place, every ant is like a cog in a giant living machine, gathering or processing food, each member performing a vital function for the good of the group.

The human capacity to think would soon undermine such a smooth running machine. People would grumble, grow unhappy in their work, or become alienated from their neighbors and wish to move along. Our capacity for independent thought can turn our society into a dysfunctional mélange of grumbling, angry, and uncooperative members. Our own intelligence could be our downfall, as people see themselves as too good or too smart to perform jobs that are necessary and an important part of our system. Perhaps these people should be given incentives to do what others are unwilling to do, but that produces more arguing. The uppity muckity-mucks despise them, and don't think they should earn or get extra pay.

We lack the instinct to follow a predestined path and must carefully learn how to live our lives. Humanity faces the daily struggle of good versus evil. We are intelligent, but also so foolish we need the grace and wisdom of our creator to give our lives a purpose. The ants don't grumble when they don't want to collect food or heroically sacrifice themselves for the good of the group.

If humans tried to live like the ants, we need the strength and wisdom of our creator; God is so important in our lives that we are unable to live without him. Sadly, many people do not know God; they ignore God, try to live without Him, or do not believe in his existence. God gave us intelligence. He also gave us spirit. These gifts are precious and incredibly powerful abilities. We need to learn how to use them, and the only way to do that is study the word of God. He created us out of love, and placed us higher than all the other life forms on our Earth. Some creatures have intelligence. Some communicate with each other. But man is the only animal aware of good and evil; man is the only animal able to understand his death, his birth and ponder the meaning of his existence. Those abilities alone should be proof positive that we are the creations of a living, loving and omnipotent God.

Arrogance, selfishness, and pride turn man into a selfish solitary brute that thinks nothing of stealing from his fellow man.
Man is the only animal that will kill his own kind for no other reason than hate or prejudice...instead of sharing, he'd kill to have it all for himself.

When God offers man eternal life, the very concept is so hard for him to grasp that he often continues in an unrighteous life, trading the love and joy of Heaven for the fleeting pleasures of this transient and perishable world. Man made joy lasts for minutes, but God given joy is eternal. Why is that so difficult to understand? Because we suffer, something that is always unpleasant and usually avoided, if possible.
The suffering we endure lets us know why evil is to be avoided. True joy and happiness are the benefits of God, while anger and hatred are the emotions of evil, the province of Satan.
We will always suffer on this world. Would it not be better to suffer and know that our suffering earns the respect of our Lord? He lived in the flesh, suffered in the flesh, and died in the flesh. He knows what suffering does to us and offers us his grace and strength to help us if we follow his advice and live a righteous life.

When someone knows his or her life is full of sinful and detestable behaviour, they often hate what they have become, thinking they are beyond help. Their vile and disrespectful lives produce guilt and shame, but they chose to ignore it and continue in their disrespectful behaviour, sometimes without knowing they can be forgiven.

Knowing you can be forgiven and continuing to suffer without hope makes no sense; if there is a chance you can start over in a righteous life, do so and thank the grace of God, for blessed is the man against whom the Lord counts no iniquity.

If someone would just read the Bible, there is no logical way to refute the blessed gifts our Lord and Savior have for us. Many are the sorrows of the wicked, but steadfast love surrounds the one who trusts in the Lord.

But changing how we live is not easy.

If someone is mean-spirited from spiritual neglect, it is hard for him to suddenly jump for joy. Understanding what God has done for us requires a lot more thought than it might seem. Then again, people are sometimes too busy with life to stop and think about what really matters, and merely drag themselves back and forth to work, always pursuing a larger paycheck…more money to buy useless luxuries, toys, or items that they love.

Nostradamus and other prophets spoke of a horrible world, full of misery, famine, evil, and war. We have the power to choose a different destiny. We can either unite under a common banner, or destroy ourselves through greed, selfishness, and ambivalence. I hope we can open our eyes in time; I hope love can spread across our borders, unite us as a common race, and give our children, grandchildren and great-grandchildren a world to live in. Right now, personal hubris and passionate hatred will kill us all. Who gave us the right to destroy something we only borrow for a lifetime. If we remember the future, realize we are only caretakers of this world; humanity might have a chance to develop into something we can be proud of.

As it is, all I see is corporate greed, global destruction, and a generation of leaders that only think of themselves. If that is all humanity will ever become, we deserve the destruction we have are experiencing. Other lifeforms make sure every single member of the community is doing well, has a job, and is a structured part of their society. Ants don't have dictators, unions, or religious hatred; they also murder their own kind, yearn for a better existence, or become jealous of his fellow ants. Like God's plan for us, they all get along, do their jobs, and continue an existence that runs smoothly and successfully. If the ant can achieve all this without intelligence and other God-given talents, how much more could man accomplish if he focused on unity and the successful integration of each member of their society. We could end our self-imposed designations of the third world. We all like to trumpet that man was both free and all men are born equal, but words are unlike the world we describe with them.
Humanity ignores the less fortunate; the rich and powerful are happy as long as they are doing well.

If other worlds turn out to have other life on them, it would only be biological evolution, such as trees or other simple forms of evolution.

Day 31

Everything in our modern world now revolves around money; be it personal gain, better jobs with higher pay, and better contracts that will also produce money. People only care about their own best interests, or the welfare of their families and friends. It really comes down to whom you know, and not how well you can do something. Unbelievably, the great American marketing machine can take an unknown and turn them into a star by following a successful blueprint of exposure and advertising.

Modern human interaction boils down to a simple formula.

Moe has something to sell. Larry would probably buy what Moe has, so Moe uses Curly to find more people like Larry. Curly charges Moe for the introduction, Moe makes money by selling his product, and Larry gets to put whatever Moe has on the shelf in his living room. While a simplistic analysis of modern marketing, advertising, then supply and demand, it's easy to see that knowing the right people can make that sort of economy work.

Sadly, it's not what you know, but whom you know.

This mentality even governs the world of art. An artist creates something, finds an agent, studio, or broker, and then hopes to get paid a lot of money.

In the art world, a high-end gallery would put on a showing of the artist's work, and if the gallery is exclusive enough to bring in very rich art lovers, the artist could command over five figures for a two-color canvass with slightly unusual graphics applied by either a putty knife or stiff bristled brush. I can say this with certainty because I am also an artist: I've created stunning oils, emotive watercolours, plus symmetrical pottery and accurate reproductions of ancient statues and busts. I haven't broken the four figure bracket, but I've made some money because someone liked what I created. Constructing something with your hands, and having someone like it enough to give you money is a nice, but strange feeling, especially if you don't value what you made as much as the buyer.

People pay you to do something they cannot do. I have a great long list of things I can do, but I would like to stress that I am trying to expand my talents. So, if you are a Hollywood producer that would like a refreshingly new script or an idea for a new show, please give me a call. All I ask for is a chance. If you don't like it, you can always call one of the thousands of professional screenwriters infesting the Hollywood area. I can also do nice birthday, greeting, or special occasion cards.

Have a look at some of the sample writing on my web site, my other book, and decide if you would like to help me continue writing. I don't know a lot of people, and I have only my writing to offer in exchange. I'd be happy to send you a signed poem or first paragraph for monetary assistance. It might be a gamble, but it could be worth money one day if I ever met the right people and have my writing successfully published.

Everything I create is raw…right from my fingertips, and without the polish of an editor or a proofreader. Ergo, mistakes and typos sometimes creep into my creations, so excuse please, so sorry…my keyboard sticks. I'd buy a new one, but I only replace something when it ceases to work…like a catastrophic failure, or something like that. Buying the latest and newest always seemed like a waste of good money…accordingly, I've never had a new car, as older ones that still run are adequate to get me from A to B. I see new or refurbished classics on the road and wonder…do they ever park in a large shopping center, and return to see new scratches and dents in that ten grand paint job?

Day 32:

How much is the doggie in the window?

Story starts: A Study in Secrets: (short story)

The stately, tree lined lane glistened in the early morning rain. It was still dark, and any house lights were strangely magnified in the mist. The deserted street seemed welcoming, friendly, and inviting. This was small-town U.S.A. Everybody existed in close proximity; pretence and toleration were common, and relations were often strained.

They knew each other's habits and everyone's routines. There was no privacy here. The prying eyes of fellow homeowners sought out every flaw, each blemish. Nothing was overlooked; nothing was forgotten. People lived with their secrets scattered up and down the quiet, residential streets. Whispers were traded, gossip was shared, and privacy needed more than tall fences. Everyone had skeletons in their closets, and nosy neighbors were He knew everyone on the street talked about him, but screw them all—he was leaving. His sixteenth birthday was last month; ever since, he'd been planning and preparing for this trip. His rich and drunken mother was an excellent source of money...she never counted it, and if she ran out, a quick trip to the bank was meaningless. He'd "borrowed" over two grand, enough to give him a head start on a new life.

The front door to one of the houses opened. Daniel Christie walked down the front walk, wearing a full-length black raincoat. His long auburn hair was tied back in ponytail, revealing a thin, clean-shaven face with delicate, feminine features. Small diamond earrings sparkled as he turned his head. The tiny white running shoes he wore made a squishy sound on the wet pavement. A large black gym bag hung over his shoulder, and swayed slightly as he walked. A car drove by slowly, it's headlights veiled in the soft rain. Daniel liked the rain. It was soothing. It seemed to swallow you up and protect you in foggy blanket. He felt safe. By the time anyone was up and about, he would be on the bus, leaving all the abuse and teasing behind. He could finally be who he really wanted to be - who he really was. The discreet anonymity of a big city would be perfect for someone like him.

As he approached the bus depot, he looked around at the town he had grown up in. There were no good memories here, just pain and misunderstanding. After his last look around, he deliberately turned his back on his recollections, and pushed open the depot's door, ready to start a new life somewhere else. Looking around the large expanse of empty seats and ticket stands, he spied a young girl dressed in black leather and jeans, stretched out across several seats trying to sleep. A single ticket booth had lights on. A bored and lonely looking attendant sat reading a newspaper.

Daniel looked at the clock. He had forty minutes before his bus departed. Steeling himself with resolve, he walked towards the washrooms and then stopped. Glancing between the two doors, he sighed and opened the woman's door. Choosing the last stall, he quickly shut the door and began to change. Several minutes later, an attractive woman in a black spandex miniskirt left the stall, and walked over to the sinks and mirrors. He opened a makeup bag, and went to work on his face: applying cover-up, eye shadow, and lipstick. He was now a she. He removed the hair-band, shook out the haircut he gave performed last night, then brushed his reddish auburn hair into a style and then sprayed it into place. Taking a last, critical look in the mirror, the door opened and another woman walked in. Daniel froze. She smiled at him, wished him a good morning, and then used one of the stalls. Daniel relaxed; as far as appearances go, he looked like a woman…an attractive woman. It was time to start acting like one. He left the washroom, and walked towards the ticket booth, an attractive woman in a black skirt. Daniel was left behind in that washroom; he was now the person he always dreamt of becoming. From now one, Daniel was Danielle. As the ring of high heels echoed across the mezzanine, the attendant looked up and smiled. Shaking his long auburn hair,

Danielle stopped a foot from the counter, met his attentive gaze, and returned the smile. With obvious interest, the man asked if he could help. Summoning his courage, Danielle replied in a soft voice, "Yes please, a one-way to New York."

"No problem," said the man, punching his keyboard for schedules, "You want the six o'clock?"

"That'll be great," said Danielle, now more sure of how he sounded.

The man started typing, and asked for her name.

Danielle hoped he wouldn't ask for I.D., and said, "Danielle Christie," pleased with the way it rolled off his lips.

A series of fast clicks came from his computer, and the printer started spitting out a ticket. Ripping off what seemed like a lot of tickets, the man said, "That's ninety-two bucks."

Danielle fished around in a small shoulder-strap purse, and handed him the money.

Boredom and routine returned to his voice after he noted the new passenger wasn't very talkative: "Thanks Miss Christie, here's your change. I hope you have a nice trip. The bus should be here in twenty minutes at Bay Four."

Danielle thanked him, put the tickets in his purse, and looked for a seat to wait. He couldn't wait to get out of South Carolina—narrow mindedness was rampant, and anyone of questionable gender was an instant target. Ever since puberty left him behind, he'd endured numerous encounters where people thought he was female. Now sixteen, all it took was minimal make-up and female attire to present a gender people recognized. Unfortunately, the people he grew up with weren't too understanding about gender confusion.

Danielle shouldered her gym bag and walked over to the waiting area. The sleeping girl was now awake; she stood up and stretched, and then fished out a water bottle from a small shoulder bag…assorted junk food stuck out the open top. It looked like they were the only passengers for the six o'clock. Danielle smiled, sat at the end of the row, and said, "Hi, nice to see someone else up at this godforsaken hour."

Returning the smile, the girl replied, "Hi back…can't sleep on these stupid plastic seats, so I'm up and at 'em. Besides, the New York bus should be here soon."

Danielle looked at the hard plastic seats and said, "Yeah, they don't look too comfortable. I hope the bus seats are better."

"Yes and no…you can stretch out here, but at least the bus seats have padding. If it's not crowded, you can use two to lay across. Bus seats are too narrow, and you really need two for any comfort. I've been traveling for two days now—I'm stiff, sore, and my bones crack if I stretch. I'm Laura, by the way. So, were are you heading?"

Danielle liked her easy-going manner, and replied, "I'm Danielle, and I'm heading off to New York City, I think. That's where my ticket leads, although I don't know how well thing will turn out."

Laura laughed, and said, "I see, one of those get me the hell out of here, take me anywhere sort of moves? I don't even know the name of this place, but it doesn't look too memorable…typical small Southern town."

"That sums it up pretty well," said Danielle, "this little backwater hole is called Onecalp. Stupid name, stupid town."

Still chuckling, Laura said, "Never heard of it, and I doubt I'll remember…the bus zips through some place, and I've forgotten it before I even see a sign. I'm heading to New York myself, so maybe we can buddy up. It's the shits traveling by yourself—I left the Florida Keys, and had this jerk sit beside all the way from Jacksonville to here. I'd move, and he'd move back beside me…couldn't take a hint. He talked about his ex-wife the whole time, stressing the ex part, and then suggested we get a hotel room. I had to get that ticket guy over there to tell him I was taking another bus, and he shouldn't be on it or he'll call the cops. I think he finally clued in…I've had to wait all night for another bus, but it's worth it to get rid of him."

Danielle thought about how he'd have to deal with guys, something that would be a problem, as he still liked girls. He noticed Laura was close to his age, and seemed like she'd make a good friend.

Laura said, "Hey, do you want to smoke a joint before the bus gets here?"

Danielle thought for a moment, and replied, "Thanks, but it's a bit early for me, but go ahead if you want."

"Yeah, you're right," said Laura, it's too early and a joint would just knock me out. It's a good idea to keep your wits about you when traveling. It's just so boring…these stopovers happen every six hours or so, and they seem to all blend together. They pull into Jerkwater U.S.A, advise you to grab some food, and tell you the bus is leaving in twenty minutes. The really big stops, like dinner, are a half hour. If you don't stick with the bus, you can get left behind, your luggage gone. These guys that got on in Florida were partying heavily, and when the bus pulled in for a dinner stop, they tried to make it to a liquor store and back. After twenty-five minutes, everyone else was aboard, so the bus driver pulled out and left them high and dry. On the road, the driver's like the Captain of a ship…they can even pull over and order you off."

"Wow," said Danielle, imagining getting left in the middle of nowhere, "that would be a definite bummer."

Always try and make the driver like you. "Hey, did you bring any food with you?"

"I got some apples, some trail mix, and some potato chips for snacks…I thought I'd eat along the way."

"Are you crazy," groaned Laura, "like eat the Depot food? That's scary…you should see the crap they have in those warming tubs. You'll turn into 'Depot man," she joked, laughing at her comment.

Slightly puzzled, Danielle asked, "Depot man?"

"You know, like in that movie…Repo man, ever seen it?"

Danielle relaxed, laughed, and said, "Right...that was a great flick. Emilio Estevez was great...I remember that part at the end...so surreal, driving that car off into the sky."

"That was cool," said Laura, "Using movie speak is something I always use in conversation, as most people have seen the movie I'm referring to, and it's a good visual way to get your point across. I was studying to be a camera operator in Key West before my life exploded."

See short story, "A Study in Secrets," if you want to know the ending, but as it is, you can add whatever ending you dream up. I recall a writing contest that begins with a well-known author starting a story, and the contest is to add the best ending.

I didn't like the way he started the story, so I wrote a version that started with his story, and then had a writer rip it out of his typewriter in frustration, as it was a story within the story...he regretted his opening, and then I wrote an entirely different story. I thought it was an original idea, but I never entered the contest. You had to pay $25 bucks to enter, a rip-off for writers and a money maker for the author. And the prize was really lame. I think you won one of those "writing packages" that self-publishing companies advertise like crazy...I get e-mails from them all the time, but they know I write.

Another example of the strange twist on publication in the Internet age, an era when anyone can post a blog, use Pidgin English, and run around telling everyone you're a published author, as you're blog is up for anyone to read.

Some of these packages include an editor's help, proofreading, a few free copies of your book, and a weird marketing campaign they say will help sell your book.

Nonsense. All that for $1,500...heck, I can get a half-decent extremely well-used car for that. With libraries, Amazon, and a plethora of e-books, people don't want to pay to read anything anymore...like Napster, they all want music for free, books for free, and complimentary magazines. The days of dime-store novels are over, and every author I've heard of has a digital version of their book up on Amazon, which seems to be the way of the future. You can get an e-reader with a small library pre-loaded for under a hundred bucks. Yepper, the days of the paperback may be numbered, but some people don't like e-books, myself included.

I like holding a real book in my hands...something I can dog-ear to mark my place, and give my eyes a vacation from computer screens.

Yes, if we were meant to read e-books, God would have given us a USB port somewhere...sorry, an old reference to flying; people used to say if we were meant to fly, we'd have wings. That must have been a long time ago. It's amazing to think of all the neat stuff we take for granted, wasn't even dreamt of fifty years ago. How did they manage? No X-box...no Internet—Wild West Days...right Pilgrim?

Note: Reality shows are much better when they have a bumbling idiot that is so stupid, you think it's an act...then you realize that's their reality. The world has so many strange and incredibly weird people, just walking around and watching others is more entertaining than any new movie, and it's free. Every time you leave your house and head out into the world, serendipity follows, insanity appears, and lunacy follows.

This guy gets pulled over by a cop.

The cop says, "You drinking?"

The guy says, "You buying?"

They both laugh...and laugh.

Now the guy needs bail money.

Overture, curtain lights, this is it, the night of nights...

We need you Superman…where are you? Batman will do in a pinch, but the bat light is broken. It looks like we'll have to make our own dreams now, and enforce them so they will get an even chance.

So many words, and their ultimate usage, make writing a dynamic and exciting method of communication. Describing a glorious sunset dripping with colours, or weaving a mystery of human treachery and cunning can be uniquely arranged to reflect either the writer's excitement, or explore the many layers of conflicting and exasperating emotions that make up a character's inner motives. Thankfully, there is no shortage of verbs, adverbs, nouns, and what have you, as the complete Oxford English Dictionary comprises 13 massive volumes. The "shorter" version is five inches thick, but now I've moved to the "pocket" edition; every page is crammed with words, and even that is an inch thick. English grows as new fads and inventions change our lives; many words are added to an already bloated dictionary every year.

Day 33:

Make sure you use sun block today, the nukes are flying.

Duex Ex Machina

The cold sidewalk bit through my thin jeans, freezing my ass. I tried to pull my coat down to sit on but it was too short. My hands were freezing, and I left my gloves at home. I started to laugh. Home…what a concept. Home had been a seedy hotel room, the only thing not bolted down a Gideon's Bible in the bedside drawer. That was last night. This night I was on the street. Alone…stuck in some strange town I didn't even know was on the map. All my luggage was on some bus, headed somewhere away from me. I'd cleverly sewn $335,000 into the lining, thinking it would get past customs. Maybe it did, although I hadn't. The last stop was a 20-minute bathroom break. Enough time for me to sprint a few blocks to the liquor store, and get more Vodka. I returned, out of breath, just in time to see my bus pulling away. I couldn't blame the driver. They always said stay with the bus, and don't wander off. That's what happens when you wander. I spent my last pocket money last night: a burger combo and a no name hotel. Now hung over, broke, lost, and at my wit's end, there I was, freezing my ass off, watching strangers pass by. Some guy tossed me a quarter. I don't know if it was a joke, or if he thought it'd help. Yeah, maybe 5 minutes of parking. After blowing on my hands for warmth, I shoved them back in my pockets. Hell, I'd left my good coat on the bus, and all my warm clothes were in my luggage. To say I thought I was fucked would be a compliment. All I could do was watch the happy pedestrians stroll by, all heading somewhere, none of them daring to "wander" off.

Convinced I'd had enough of the pity party, I got up. At least if I walked around, it would give me some exercise and help me feel warm. I crossed some street then headed down another. I had no idea where to go, so I just walked. Jamming my hands in my jeans, I felt some change and a bill. Pulling it out, I saw how rich I was. $6.76...hardly enough for a local bus. Trying to think, I continued to walk, hoping some brilliant idea would pop into my head, or some duex ex machina, divine intervention, would pick me up and put me back on my bus. Shit. All that money...just motoring away, heading to Las Vegas...the city of dreams. I wondered if I talked to the bus company, they might return my luggage. Right. Twice across the border. Some border agent might start poking at it, wondering why the back felt so dense. Thirty-three packs of $10,000, and one pack of 50 $100 dollar bills. I'd be happy with one. I turned off the dismal street I was on, and saw the flashing lights of a casino. I headed over. I'd either lose all my money, or maybe catch a break.

Six hours later, my hair slicked back with tension sweat, I swaggered out of that casino. My 5 bucks became $10, then $200, then and $10,000. I kept at it...blackjack, then craps, and finally roulette. Remembering my frozen ass and all the despair in the world, I put $10,000 on red 17, winning big time. I then bet odd, bet even, and bet red. They were almost pouring drinks over me, trying to get me drunk and reckless. I guess I was reckless, but everytime I stopped to think, I remembered that frozen sidewalk and said fuck it...I could only win or lose, so I went for it...and won. That $335,000 was a distant memory. Every pocket I had was stuffed with $100 dollar bills, with a huge wad in my underwear, and a big paper bag. $1,400,000 bucks takes up a lot of room. I was piss drunk by this time, but I knew after that last roulette spin, going from $700,000 to $1,400,000, it was time to quit.

Just after they gave me the $700,000, I paused for a moment. That was a lot of money. It was more than I had zooming around on some bus. I looked the croupier in the eye, and looked at the manager that had wandered over, curious about all the action. I could see fear in their eyes, and I remembered my frozen ass sitting on that pavement. That's when I knew I'd double up...either that, or go back to my spot...it might still be warm. I remembered the bus driver warning us not to go "wandering" off. When you have fuck all to begin with, and don't give a fuck, you wander off...I'd wandered off, big time. When it hit red, I looked at the manager; he had that look of despair I wore not too long ago, and just like that, I'd won over a million bucks. Like wandering buses on the highways, things come and things go. My bus left, but my ship came in, and I didn't have to swim out to catch her. Of course, they wanted me to have a room and stay. That's what they all want...a chance for you to get stupid and lose it all again. I turned into the first clothing store I found, bought a long, Burberry style coat, a red scarf, dress shirts, pants, shoes, and a nice attaché case. I think I freaked the guy out after I paid for everything. After getting properly dressed, I opened the case and emptied my huge sack of money into the case. He looked at me, and I just said, "Casino...I lucked out." I gave him a hundred dollar tip, and he was obsessively obsequious, asking if he could do anything for me. I wondered where all that help was while I sat, freezing my ass off, after "wandering" off. I left quickly, before I punched him.

I got some directions to a nice hotel and left. While I was reading street signs and trying to find my way, I saw a guy sitting on the street, shivering in the cold, a pathetic Styrofoam cup in front of him. He was about my age, with an inadequate wardrobe, and even looked like me...same length of hair, same color, and similar previous wardrobe. I broke out laughing. He gave me a strange look, and I reached in my pocket, found a wad of cash, and handed it to him...about 3-5 grand. His mouth opened in surprise, and I said, "Duex ex machina baby...you never know when luck walks by...go get yourself a warm coat." He stammered out a thank you, and I just said, "Hey, don't worry about it...money...it comes and it goes. You want to go get something to eat and have a few drinks?" He jumped up, broke out in a big grin, and said, "I don't usually beg, but you wouldn't believe what happened to me." I couldn't stop laughing.

Day 34:
Sunup was at 5:35 this morning.

This continues to be like a written recipe: a little of this, a dash of that, mix on puree, bake for twenty minutes, then ice. Question: if you randomly opened the dictionary and hit the first word you saw, how long would it take before you had a sentence? (a reference to the psychologist who said tons of monkeys on many typewriters would eventually produce the works of Shakespeare—never believed that one)

Coyame UFO crash in Mexico…the Mexicans retrieve UFO, the C.I.A. sends an intercept team, and find four soldiers dead. The Mexican let the U.S. take the craft, and what happens? A cover-up. Why did the soldiers die? Alien radiation or disease…no one will ever know until the U.S. comes clean…don't hold your breath. Back to Earth.

My poor used car was an innocent victim of crime. Vandals smashed the passenger's window, then the driver's, slashed tires, broke tail-lights, and scratched the finish.
I tried putting a pity note on it:
Please don't hurt my car; it's a nice car, and even goes out of its way to avoid other vehicles, and it doesn't hit other cars that yell bleeped out obscenities when minding its own business. This car's been in four faultless accidents, and has been to hell and back. It has nightmares about the heavy metal car crusher in the sky. It's so paranoid, if I play heavy metal music, it starts to shake all over. Thank you..
Meaningless acts of destruction sometimes stop if you add meaning; either that, or the *%#@% little pricks moved.

Much ado about nothing…the Winter Olympics came to town.2010.

Olympic gold for $$$security:

One billion dollars, give or take a few million: does that amount of money ensure the 2010 Olympics will be safe from terrorist threats? Do the organizers believe the crowds will feel safer knowing tons of money has been wasted so those professional security guys can practice radio protocols? "Alpha one this is Tango three, do you copy? Any bad guys around? No? Damn, well, I'll check you later. Out." Of course not: it merely ensures the security companies will enjoy a big payday, but that is the only fact we know for sure. Al Qaeda & Co. must be chuckling.

When analyzed from any objective standpoint, there are a lot of "undisclosed" facts. These "leads" or "dangers" are not open to debate, are supplied by self-proclaimed "experts," and utilize such standard and out-of-date methods they wouldn't deter a mouse from a block of cheese. First of all, let's look at the cost of this style in security…I'm no expert, but I image there are a lot of different strategies involved, just like anything else. I'd be a bit concerned about a man with a rail-thin face but a gigantic gut, for one, or a guy looking entirely out of place…like a guy that's never dressed for snow before, and has brand new everything, along with a massive puffy coat. Then there's the old shifty look. What did they use? Manpower. Hiring a slew of carbon-copy police and security agents/clones, carefully placed in what they consider trouble spots, would always entail a hefty, per-hour, bottom-line. Unless he's a stealthy suicide bomber, not the kind that has "terrorist" stamped all over them, some of these guys could blend into crowds, and no one would know. Or, they would stand out as much as the undercover security guys…the ones that try to look "normal," but end up looking like they were all holding "I'm a cop, pretend I'm not here" signs.

Perhaps we can break it down into a list of what our special "extra protection" police costs would be, and compare them to what we now list as "Olympic" costs:

(These are estimates, add up or down, it still provides perspective)

Mobilization of all "in-place" security forces:

Firemen	200.000
Ambulance stand-bys	200.000
Army/militia, integrated	400.000
Integrated Police Forces/auxiliary units	free
Volunteers (who get to wear cool coats)	free
Re-activation of retired and disabled forces	200.000
Rental of helicopters and stand-by airplanes	<u>500.000</u>
	2,200.000

2.2 mil seems a bargain if you use what you have…and what are we being protected from? Overall mental stress? Shock? Loss? Surprise?

Compare that amount with the billion they are currently suggesting they need. Police and special terrorist units are the first to admit they do not know how to combat terrorism. No one does, not even the terrorists. They use our greatest weakness against us, and it seems to work: trust. Obversely, you can't go around and search everyone…that basically means the terrorist have achieved what they want: making us feel unsafe in our own country.

Given that, no amount of security in the world would prevent a suicide bomber. These are intelligent, well-trained combatants. They would wear a specially designed vest with pockets for as much C-4 as they could carry. Cold weather makes it even harder to spot them. They would just integrate themselves in a large crowd and explode. Thousands of police officers, helicopters, and jets can do nothing to prevent that. In fact, the more attention we create, the more likely we could become a target for terrorists. Having two cops on every corner does nothing more than give the cops a buddy to chat with and distract him, and keep him company.

We all know one cop would be more alert, as that's all he has to do, but they always appear in pairs. And, that is exactly what will happen. A normal police presence would be sufficient; they would be there to call for additional emergency vehicles, render aid, and, by their very presence, act as a deterrent against zealous and disruptive protesters, while making it known to any terrorists that there is indeed, a police presence. They will naturally assume there is a huge force of undercover officers already in place. Why not capitalize on that presumption and use it against them? Sun Tzu, in his "Art of War," declared the best defense is to use the enemy's natural fears and assumptions of defensive positions; that leaves you free to focus on your offense. A response time of five minutes can be accomplished with five hundred men: five thousand becomes cumbersome and a massive hindrance. But, the more high-paying jobs they create, the better the security people feel about it. This is an honest assessment: people need to face the facts.
Anonymously yours,
The cost-effective man and free advice.
(Post note: nothing happened)

This would be good, old-fashioned policing; combining forces with RCMP detachments, CSIS, and the Canadian Armed Forces shouldn't be overly complicated, expensive, and a huge cost to the Canadian taxpayer. What do we pay these guys for in the first place? When they are actually asked to do something important, they scream they need more manpower and, of course, more money. In the real business world, we have an attitude for this: suck it up and do your job. After years of sitting around on your duff, it's time to step into the limelight and let the world see how good you are. A lot of people are volunteering and doing Olympic jobs for free (if they get to wear those cool coats). The security business is now a huge drain on capital. And, they don't seem to do more than stand there with a high tech earpiece, gun, and cool 3G smart phones.

Day 35:

Philosophy is what explains personal eccentricities.

Life. A short word with countless forms, questions, and metaphysical complexities, it is a term with endless possibilities. Social and personal evolution, complex organic biogenesis, life itself creates such convoluted questions that all answers will be debated, discussed, and dissected but never agreed upon. Our own origins are a philosophical, scientific, and spiritual conundrum that will never have an acceptable answer. The possibility that several civilizations, equaling or surpassing our own, could have existed on Earth…societies that were forever obliterated, misted over by disaster, their memories now past myth, a mere shadow of what might have been. Recent archeological discoveries, inexplicable and unknown ruins, have created questions previous historians ignored, or glossed over as they upset their perfect theory of historical events. Sunken blocks off the coast of Japan, massive in size and geometrically shaped by man, Gobekli Tepe, Nevali Cori, the precisely carved megaliths

Scientific or organic knowledge of life is a small part of its overall meaning. Learning to live together, create a functioning macrocosm, is another aspect of life that we might never achieve, destroying ourselves in our own selfish belief of strength and power.

Now that we have evolved into similar minded beings, individual rights, freedoms, and goals will forever be problems we must contend with.

This aspect of our own awareness creates the very problems that hinder people from combining into a well-functioning whole. Envy, jealousy, and a host of other human emotions are always going to present problems. Spiritual awareness offers one of the only possibilities for enlightenment and peace.

The question can become whether to work within the system, push yourself to the top by sheer willpower and strength, or give up, and admit defeat. Whether a person chooses drugs and life outside of the system, or struggles endlessly to carve out a niche for themselves within the system depends on personal beliefs and our personal source of strength.

I choose to believe that despite the overflowing welfare lines, there is a God that loves each and every one of us, who one day, long ago, blessed us by allowing His own Son to die at the hand of sinners in order to cover the ugly nature of evil, and give us something to hope in that will allow us to shed our sin, and bask in the divine grace of His forgiveness.

For what is this Earth but a collection of microscopic particles, all interacting in a miraculous design to continue the cycle of life. I believe that method is so magical it is a miracle, and only God performs miracles. Life itself is a miracle…and ergo, a creation of God. This may be a simple rationalization, but once stripped of complicated theories and guesswork, our very existence comes down to this simplification. Just as there is light and darkness, life and death, male and female, solid and gaseous, naked or clothed, so there is good and evil, the very complex balance that is of such importance to our living God. We can see certain balances with our eyes, but underneath existence is a vastly intertwined network of invisible fields, nanoparticles, and even extra dimensional factors that are all combined into a magnificently successful system. We cannot see the wind, but we can feel it: we cannot see gravity, but can witness its effects…once again, we can understand the multidinous factors that are summed up in the word life.

Given all these unseen forces that are mandatory to produce universal counter balances and substance, the question of life can enter a more sophisticated form of analysis, a form that questions why there is life and death. We understand that death occurs when the body ceases to function, but what is the magical spark that begins life in the first place?

The only answer is our soul: our spirit or life-force, the miracle that forms our personalities, our intelligence, and our emotional phases. Just as there is joy and sorrow, our emotions seem to exist in a balanced form…the balance that seems to affect live throughout the universe. It can be argued that our souls are the emotional indicators that maintain our spiritual balance…an invisible, ineffable part of our existence that cannot be seen, measured, or dissected in a lab. Just as we accept invisible forces as real, so must we understand that our spirits and individual personalities survive by an unseen entity that is our soul. When this fact is accepted as real, humanity can also appreciate that the balance found throughout every force in the universe must also exist in the soul…in the nature of good and evil. Our ability to choose freely is an important factor when we examine the balance that must exist in the soul. Ultimately, the fact that we can choose between good and evil, proves the importance of our main reference to this phenomenon, the Bible. The Bible is unique and will always be a source of debate—for some, it represents the Word of our living God—others freely choose to believe the Book is whatever they believe. Nietzsche said God is dead: that statement is baffling, because it admits the existence of an omnipotent God, and through that admission, creates a question we mere mortals are unable to answer. Perhaps the statement that God works in mysterious ways is the best expression to prove that God is very much alive, and His inscrutable ways can only be understood by believing and reading the Bible.

DAY 36:

Post-Structuralist Philosophy and the Reality of Imposing Order

The title of this document suggests a well reasoned and thoughtful opinion on philosophy...anything else would be considered a personal attitude towards something, or an attempt to entertain a reader by creating a series of enjoyable words, a creation that would be considered art or creative writing. Assuming these words convey meaning is another area of human thought that requires either scientific proof, or an agreement by the reader that they are forced to think while reading this article. These previous statements also fall into the umbrella-like structure of philosophy: accordingly, this essay is a productive document on the nature and meaning of philosophy. If a reader can imagine bedrock, or the absolute bottom rung on an intellectual ladder, anything above that would be considered progress...unless this had all been said or written about before, making this entire document redundant or a waste of effort. However, that style of thinking will never rise above bedrock, and intellectual progress is therefore something we can never achieve.

Philosophy has always been associated with studying the meaning of life, and our ultimate ability to understand our place in the world. It is concerned with the nature of human intelligence, the limitations of this ability, and the potential it has to truly understand the meaning of life. There are many sub-divisions within philosophy, off-shoots that focus on specific areas of study, trying to impose meaning and structure to random speculation. Accordingly, attributing meaning to the vast expanse of life is philosophy's purpose, however vague and imprecise that basic meaning is. Some branches of this thinking science even suggests that the struggle to find meaning in things we don't understand is an impossibility, an argument that shows how elusive true understanding can be.

Essentially, philosophy can be described as studying human understanding and every aspect of that gift, from a negative to a positive analysis of thought. Existential thought strips away acquired knowledge to extract a simplified form of life, establishing awareness from ignorance, simply by believing that we exist because we think we exist. Thought is the only tool scientists use in philosophy…since writing is a method of recording thoughts, therefore thoughts exist as writing…according to Logic, a sub-section of philosophy, specifically proved by a syllogism, or progression of knowable truths. This form of Logic also exists as Empirical thought, or something that is concluded by a series of knowable and verifiable truths. Even a simple discussion of philosophy must explain all underlying truths, otherwise is would be an unknowable form of writing known as pure fiction…but when fiction becomes fact, it is elevated to a provable result, something imperitive when using any scientific method. Without this proof, experiments are relegated to goofing around, and any written record of this becomes a mere story or fictional document of guesswork. Since that would be pointless, this essay endeavors to attain provable points…otherwise, this entire exercise in words would be worthless, in a scientific sense. This statement is based on an observation from arguably the wisest human to exist, the great Solomon the Wise. Jesus would be the wisest person to ever walk this world, but because He is from Heaven, it's not fair to include His omniscient capability.

Structuralist thinking is basically humanity's attempt to impose meaning on a seemingly random or pointless occurrence. Starting with microcosmic structures, imposition of order begins assigning meaning to the very small and progress to the macrocosmic, or very large. Accordingly, creating laws to define small ecosystems or phenomenon is much easier than imposing meaning on complex ecosystems…systems that include human behavior.

DAY 37:
What?
Good for Something

Congratulations. You've just made the wisest and most honorable commitment of your entire life: you've turned your life over to God. You've admitted you are a sinner and believe the message of Jesus Christ, "that he who believes in him shall have everlasting life and never perish." Now what. Accepting the truth is easy, but living a righteous life is not as easy as it sounds. People think they are too smart, and too advanced to believe in myths, and many consider the Bible a collection of myths. Parts may be simplified in order to teach a non-technological mind-set about creation, but that creation happened—it could have taken millions of years of evolution, but there is a certain point where Neanderthal man turned into Homo Sapiens, with an advanced brain. Believing that was because of alien infused DNA is the same thing as recognizing that God had a hand in our development…anything not of this Earth is extraterrestrial and alien. Ancient aliens champions show numerous examples that our ancestors saw things that were clearly otherworldly, and depicted those encounters in carvings or etchings. Heavenly technology can explain a lot, as the historical record clearly shows that some really strange events happened. Whether angels or aliens matters not, as they are basically the same thing: what is important here is to understand the messages they left—love, peace, and harmony are key to any advanced civilization. Well, entire books delve into this, so I'll just say what I've always believed: anything is possible, if it benefits God.

You feel full of life, bursting with happiness and tears of joy are running down your cheeks. You've just prayed a simple prayer that throws out your list of sins, nailing their evil actions to the cross, so the blessed forgiveness that is Jesus' blood can wash them away. You have been forgiven by your maker.

To quote his response in this situation, he said, "Go forth and sin no more." Quite the task, if you live in the real world. You can't run off and join a monastery, so you're going to have to learn how to live a good life surrounded by shadows of your former self. The liars, the cheaters, the thieves, the blasphemers, the selfish, and the greedy: in short, a world that still values everything you have renounced. Even if you have the support of a large church, you can't drag that Church around with you. You have to learn to stand on your own, make the right decisions, and learn to treat people as you wish to be treated. Basically, you have to relearn everything there is to learn about how to live.

Let's assume your inner circle-of-friends are good people, trustworthy, and full of good works, but they are not Christians. You are still brimming over with the excitement of your newfound faith. You call or text everyone on your friend list, hoping they will all drop what they are doing and call to congratulate you on joining a worthwhile cause. After a few hours, you finally get a call, but the person doesn't mention your recent conversion. John wants to know if you're going to Terry's party on Saturday. Slightly dazed, you recall the party and tell whoever called you would get back to them. You realize you never told John about your recent conversion. Time goes by and no one else calls. No one you told about the most important decision in your life wanted to call and congratulate you on choosing the right path. You suddenly realize your old friends don't know what to say to you; your decision has made you an outsider, someone that has joined an exclusive club that non-members don't understand. You slowly realize you don't share the same values as your old friends, and your old friends don't share the values you embraced with love and grace. You are now alone in this world of sins; your only friends are sinners, you don't know any Christians. You remember the Bible said to read your Bible, pray, and surround yourself with your Brothers in Christ. But you don't have any brothers in Christ, you only have your believe in Christ. You pray for guidance, and read your Bible. Do what you can, and do the best you can…what else can you do?

The day you dedicated your life to Christ is finally at an end. It's time for bed. You dutifully read your bible before falling asleep. You anticipate a wonderful tomorrow walking with the Lord. At seven o'clock, the alarm rouses you from a deep sleep. Tired and grouchy, you pull yourself out of bed and hit the shower. The apartment complex you live in ran out of hot water as the earlier risers used it all before they left for work. You honor them with a string of curses. After your shower, you head to the kitchen and get a coffee. Like a satisfied addict, you enjoy your first cigarette with your coffee and flip through the remaining sections of yesterday's newspaper you never read. You read about one of our soldiers giving their life for democracy. In hasty anger, you decide the whole lot of scumbag terrorists should be nuked. It's at this moment, as you greedily suck on your morning smoke, that you remember the decision you made to give your life to the Lord. You glance at your phone. Still no calls. Suddenly, you wonder what happened to the tingling awareness of joy you felt when you acknowledged your sins and felt a flowing sense of pure love wash over you. You stub out the cigarette and hold your head in your hands. Confusion washes over you, followed by fear. In your rapture, you announced your decision to over 40 friends on you cell phone's dialing list. Embarrassment begins to taint your prior resolution. That feeling of indescribable joy isn't there; you feel sore and cranky, and all your thoughts are punctuated with four letter words. You feel as if you might have bitten off more than you can chew. You don't understand how to live as a Christian; Church is over, and you won't have any Christian contact until next Sunday. Routine creeps in and you get dressed and run off to work.

Your life is as it always was; you want to change, but you don't know where to begin. You feel lost and alone; surrounded by temptation and your former sinful existence, you automatically slip back into your old ways. Then you remember some of the scripture you read. You want God in your life and decide that life without God is not life; therefore, every action and response takes time. You need to learn how to respond, think about what you say and act the way you know Jesus would want you to act. Slowly but surely, you learn to overcome the difficulties you face. You're not embraced by new friends, so you learn to live alone, alive in your faith and trusting in your new belief.

DAY 38:

More later

A novel exert, or a novel beginning. It does have possibilities.

Note: this is now part of Drugstore Cowboys, my first book, in some way, shape, or form.

The studio was cold and damp. Gusts of wind drove the rain against the plastic covered windowpanes with increased fury. Slight rips in the odd assortment of shopping bags covering the windows allowed the wet night inside. Puddles under the windows ran together into small rivers, coursing over the old warped floorboards until they drained into an ancient pipe hole that once connected the rusting hot water radiator. A faint trickling could be heard whenever the wind subsided.

Sean Ross was totally unconcerned with both the weather and the continuous stream of water that soaked the floor. At least his head was dry. Seated on the only piece of furniture in the room, his attention was focused on a small cotton ball held against the side of a bent spoon by a 1cc disposable syringe. Sucking the spoon's clear contents with the syringe, Sean's eyes had a dazed but intense gleam to them, as if the ritual he was performing had a sacred tradition, like the Chinese tea ceremony. Shaking slightly, he tightened a belt around his arm until a vein popped up, plunging the needle and its contents into his arm in a well-practiced motion. A warm euphoria spread throughout his body, and he settled back on the sofa, his face now glazed with a satisfied mask, his shaking and twitching body now relaxed under the narcotic's insidious influence. The world in which he existed seemed to melt into a phantasmagorical nightmare, images that had no substance for him and lacked a meaningful explanation. They were just part of his world, a world that took second place to his inner world that was now running on heroin. Yet that second place world, no matter how hard he tried to ignore its abominable reality, that world had become his life, and its impossible existence always managed to penetrate the frail narcotic shell in which he tried to hide.

It was a paradoxical life. The past brought back too many painful memories, while the future looked just as bleak. With nowhere to turn, Sean sough refuge in his drug-induced sanctuaries, utopias far removed from his present consciousness. Everything he had once known and cherished was now banished to a twilight graveyard somewhere in the depths of his once rational and productive soul. A tormented soul that was searching for something he thought was unattainable.

Like a New Years Day, apart from the pounding storm, everything was quiet and serene, lifeless and forlorn. An echo of emptiness, like a shroud of quiet desolation; a vacuum was sucking in the air, silencing the room.
The emptiness ran through his body, focusing on his depressed spirit, and anguished soul. He wanted more from life; something that would make him aware he was alive, but in his current state, all he could accept was that he was beyond repair, unwanted and destined for loneliness and despair. His friends only wanted him for the drugs he stole from the pharmacies. They didn't care whether he was laughing or crying; as long as he had what they wanted, they were happy. Sean had a memory he still remembered from when he was 8 years old; he felt warmth, love, acceptance, and joy, everything his spirit now thirsted for. It was when he felt the love of God, when the Spirit of the Lord drew him to the altar of a church where he knelt, prayed with a Pastor, and gave his life to Jesus. It was the only time he could remember true happiness. Now he felt he was beyond forgiveness; beyond reconciliation and doomed to wander the Earth like Cain, searching for God, but never able to attain the goodness and Godliness he needed to make things right. That was the main reason he used drugs: to try and inject a false sense of well being, a temporary fix of warmth. It was something he tried not to think about.

The encroaching walls were driving Sean insane. Despite the rain, he decided he needed a walk. He put on a thick hoodie and his heavy-duty raincoat and headed out. With no set destination, he began walking East on Danforth. Passing a bus stop, he saw a bus coming and decided to take a ride. The bus pulled to a stop, the large wheels passing through a huge puddle that kicked up a wave that Sean managed to avoid. He paid his fare and took a single seat, as it was nearly empty. The bus was destined for downtown; the location was fine with Sean. At least he would be with other people and have something to do. He pulled out a battered paperback and settled in for the ride.

The ride improved Sean's spirits. Foul weather had a way of making people friendlier; everyone who got on was glad to be out of the weather, and would smile and nod at the other passengers. There was an unspoken connection between fellow humans who were enduring the same torment and enjoying the momentary respite from nature's fury. After a 45-minute ride, the bus pulled into a turnaround at Richards and College. Time to embark and face whatever nature was dishing out. He started walking along College, doing up all the zippers and snaps on his raincoat, effectively keeping the rain out. With both hoods up, he was both warm and dry. The rain had let up a bit, but was still steady and wet. Wandering aimlessly, he found himself standing outside an Anglican Church somewhere on College. For no reason, he walked up and tried the front door. It was open. The old stone building was quiet and soothing. Without knowing why, he sat on one of the end pews and stared at the huge cross at the front of the Church. Forgetting how to pray, and not knowing where to even begin, he got up to leave. As he entered the alcove, he met who he assumed was the Church's Pastor.
"Good evening my son", said the Pastor," did you come to escape the rain or is there something I can help you with."

Sean looked at the friendly man. He had a presence about him that instantly made him feel at ease. Without knowing why, he decided he would ask the priest some questions, for his heart was heavy with sorrow.

"I'm sorry to disturb you Sir," began Sean, "I was walking by and something sort of lead me in here. I haven't been in a church for a long time, and it just seemed like the place I should be."

"Would you like to come in my office and talk?" invited the priest.

"Well, I'm sure you're busy," stuttered Sean, suddenly overwhelmed with emotion.

"Nonsense son, that's what I'm here for. I think there was a reason you were led here tonight, and I'd like to hear what you might have to say. That's my line of work, so don't worry about it. I sense you're going through some issues now. Perhaps I can listen; it helps to get things off your chest. I promise I won't try to baptize you or beat you with a Bible."

"Sure," said Sean, suddenly feeling an inner lightness that dashed his depression, "that sounds wonderful."

Sean followed the man with longish white hair down a corridor and into an office lined with books. A guitar sat on a stand surrounded by bongos and other percussion instruments. Unusual objects hung from the walls; it was not what Sean would consider the office of a Pastor.

"Well I'm Father Peter," began the kindly old man", why don't you tell me why you were out on a night like this and ended up here".

Sean could feel embarrassment redden his cheeks, as he knew he couldn't lie to this man. That meant explaining his drug use and inner depression. Subjects he wasn't particularly proud of telling strangers about.

"Well Father, my name is Sean, and I've had a problem with drugs for quite a while now, and I'm at a point where the drugs are not making me feel any better. In fact, I feel like I'm dying inside. This would be a confidential talk, wouldn't it be?"

"Of course. Feel free to tell me anything. I've heard some pretty incredible stories over the years, and I've witnessed some miraculous changes as well. Tell me anything you want, it won't go anywhere."

Sean relaxed and felt that this was the time to tell someone the truth about what he was going through—all of it.

"I'm known as a drug store cowboy: someone who goes into a pharmacy and takes all their narcotics. That's one of the reasons I get so involved with drugs. I sometimes have a lot of them."

The priest didn't know what to make of my story, and was short on solutions. As per usual, I left, returning to the only life I knew—a hefty shot of opiates.

Bit for a novel?

Short story or novellas—whatever, add to bits and pieces, it might come in handy.

I wonder why this never worked:

Dear Mr. Harper:

I am the son of a World War II veteran and have had an impossible time looking for a job since I graduated from a Computer Networking Degree in 2002. I also have a B.A. in English and History from York University, but find that every door I knock on is closed and someone with inside contacts eventually takes every position I apply for.

During my practicum for the Computer Networking Degree, I worked for a Native school funded by the federal government's Indian Affairs department. The school was quite pleased with my level of education and considered hiring me full time. Unfortunately, only three people were considered, and the job went to the office manager's nephew, who was just finishing a similar computer re-training course—nepotism in action.

No matter where I look for a job, I find nepotism to be a real problem and have no way to fight it because I don't have the right connections. I thought that by writing you, you would gain a sense of who I am, and based on that familiarity, might recommend me for a Federal job in British Columbia.

My Bachelor's degree proves that I am capable of learning difficult skills and would be able to function in any job with minimal training. I've also gained a lot of office experience and would benefit any department that might give me a chance.

All I want is a chance to work and show what I can do. Given the effort and time I've put into looking for a job, it would be a graceful act to lend me a hand. All I need is to get in the door, my hard work and diligence would do the rest.

Thank you for your time and consideration in this matter.

Me...see me resume, or assume the position, you fat-headed prick.

Day 39:

My inner monologue sounds like I'm a stand up comedian. Perhaps I should try Yuck-Yucks, and just say what's on my mind.

I know what perfidious is all about; being treacherous, the act of violating faith, trust, or allegiance. That's when you need to repent, but that only works if your really sorry for what you did. I recall the Apostles asked Jesus how many times they should forgive someone, and Peter, I think, said, "As many as seven times Lord?" Then Jesus said, "Not seven times, but seventy time seven." That's a lot of apologies, but what can we all do, since we are sinners, and sin on a daily basis. What really gets me is when Jesus mentioned that the mere thought of doing something makes you guilty, and that's something that happens all the time. Like that dude who was yelling at me in traffic today...I'm glad I had my music blaring, as I'm sure he wasn't telling me how much he loved me. That was hilarious, watching his angry frown and babbling lips, as I smiled and waved, oblivious to what he felt, and unable to hear over the fantastic music I was listening to. That seems to be a lesson on life. If you can ignore the nasty side of things by putting up a wall of your own inner music, you don't hear the curses and foul-mouthed expression people spout out for no apparent purpose. If we are going to be judge for behaviour like that, I think a lot of us are in deep trouble when it comes to getting into heaven. Perhaps that's where asking for forgiveness on a daily basis comes in...you recognize your failings, confess them, and hope you'll learn and become like some mystical guru, above the dirt and traffic of life, up in some dream-like world where perfection rules, and chaos is conquered.

Perfection. Having all the elements or qualities requisite to its nature or kind in absolute balance; complete. Also, the highest degree of something; without defect, flawless: it was a perfect hole-in-one. The definition of perfection, if applied to the only life form that deserves this definition, is woefully inadequate to describe God. God is beyond perfect; the creator of everything, the source of all power and the divine engine that drives his miraculous creation, the Universe. And with all that incredible intelligence and control, he made Man in his own image, hoping that Man would recognize His Supremacy, obey his Laws, worship Him and find the strength to fight the only thing God hates; sin. God made man an immensely complicated being, able to function and reason on his own, able to marvel at the wondrous splendor of God's creation, but given the source of his own downfall. Free will. Everyone knows the story: the serpent tricked Eve to eat the one fruit God forbade him or her to eat. By defying God's one request, Eve brought original sin into the Garden of Eden, enraging God and separating him from us for a very long time. Thankfully, God is immortal, omnipotent, and full of love. Our disobedience must have hurt him more than we have ever suffered. And, if you recall your history, the human race has had some pretty bad periods of suffering throughout the ages.

We are fortunate that our God is a loving God, a God that saw that his perfect creation could overcome evil, but only with a lot of willpower and resolve. After getting booted out of paradise, when man began to multiply upon the face of the Earth, we somehow forgot about Him again, and this time nearly got wiped out of existence. If it were not for one righteous man, Noah, we would have replace by mankind 2.0. When God handed us our brains, we immediately fiddled around and managed to cross-circuit common sense, engaging in sinful acts and selfish indulgences again. We didn't pay attention to the most important thing about this world: that this is GOD'S world, and we need to get down on our hands and knees and praise him for giving us the chance to enjoy this majestic globe. Thanks to Noah, and God's grace and munificence, we got another shot at figuring out how to live. But with so many different branches of people popping up, God took pity on one race and claimed them as his chosen people. It just so happened that the Egyptian's had enslaved them at the time, but all things are possible through God. So, He promised his spokesperson, Moses, that he would send so many plagues against the Egyptian's that they would be glad to get rid of them. He also promised them they would find a great place to live, and that they only have to obey Him and follow his orders. There are quite a lot of stories written about this, but that's the basic story in a nutshell.

God's chosen people, the Israelites, became witness to God's great power. Even after the Egyptian's suffered through 10 horrendous plagues, they were still too greedy to just let them go. They sent an army after them: God parted an entire sea, let his people cross, then let the sea fall on them when his people were safely across. Quite the show of strength: the Israelites must have been deeply impressed.

Anyways, they started out for the Promised Land, and God was pleased; he told his people, through intermediaries, how they should live. Everything was acceptable, and everyone got along just fine. However, that stupid gene God gave us popped up again after about 40 years of traveling from Egypt to Israel. That's quite a journey for an entire nation to make, considering they were carting everything they needed to exist along with all their earthly possessions. Our loving God had made covenant with these people. He didn't ask for much, but these people couldn't even do that. It was only through the wonderful grace of God that we didn't get wiped out for good that time. Moses was so mad he even broke the tablets that God gave him with ten laws engraved in the stone.
Ten commandments and they couldn't even follow them. Today we have more laws than people, and it takes a team of lawyers ten weeks just to find out which one you might have broken.
Anyway, back to our history. They arrive at the Promised Land and prospered for quite awhile. They have a rich history; after a few ups and downs, they

were conquered by an old superpower, the Roman Empire, and things seemed to quiet down for a bit.

After their escape from Egypt roughly 3,000 B.C., God intervened in their lives by implanting thoughts into prophets and they in turn wrote down what they were told in the Pentateuch, a series of books that contained their history, God's Laws, literature and wisdom. After they compiled this huge collection of educational and enlightening stories and teachings, things were quiet again for about 200 years. Either God had been doing some thinking, or just doing what he always knew he was going to do, but things started happening that would affect the whole world, not just the Israelites. We have no idea how God works. He's omnipotent. He exists in the past, present and future. He is so beyond us, all we can do is worship him and do everything he tells us to do so we have a chance to stay on his good side. We've figured out how some of the world works, and it is a masterpiece of engineering that we might never truly understand, or just garner a general idea of how it works.

We might explain what seemed like miracle to the ancients through our knowledge of science. But, God controls the science, the sub-atomic levels, vast gravitational forces that keep our world in orbit, and just about everything, which keeps our Universe functioning.

Day 40:

This seems to be an important number in the Bible...they were in the wilderness 40 years, Jesus was tempted for 40 days, Noah endured 40 days of rain, and so forth.

To give is to receive; presenting someone with a gift from the heart is more rewarding than a fist full of dollars. Emotions cannot be bartered as with money. It takes a lot of introspection to pinpoint exactly how you feel, but when you're happy, it's easy to know how you feel. Emotions spring from within, jump with overwhelming joy or bear the burden of sorrow; they are unpredictable, ephemeral, and fickle. Emotions, not wealth and fame, form either heartache and pain, or ecstasy and elation. Essentially, we feel good or bad; a battle that has existed since the dawn of time, thanks to one man's colossal mistake. Adam ate the only fruit he was expressly told not to eat. Curiosity kills, no doubt about it, and it continues into this day. I guess if he didn't eat that fruit, humanity would live a dull life, but it would be full of joy and a lot of fruit...minus one. What was so special about that fruit that Adam was willing to risk separation from God just to have a bite...it must have been all the lies the serpent spun, dazzling Eve with his evil salesmanship. Step right up...the miracle fruit; one bite and you'll know everything, but you'll get tossed out of paradise on your naked ass, but don't let that worry you...just imagine the juicy flavors running down your gullet. Perhaps it was the opium poppy, and Adam wanted to get high...or a huge pot plant, dripping with hash resin, crying out to be smoked. Right...I can see Adam, walking with God, saying, "You don't have any rolling papers you could create, oh great one?"
Another question...if God wants to be a father to us, how close do we become? Definitely not close enough to trade jokes...then again, they say when you are touched by true nature of God, all you want to do is serve Him in ever way possible. Then again, I've heard priests tell jokes, like the one with God golfing with Moses and Jesus. They all missed the green on the first shot, and

only God put it on dry land. Moses and Jesus put the ball in the water hazard. Moses took his club and parted the water to get his ball, and Jesus merely walked on the water to reach his ball. God had a squirrel run with his, give it to a deer, which gave it to an eagle that dropped it in the hole. Jesus turns to God and says, "It's not nice to show off dad." Is that disrespectful, or merely part of our human condition…the need to laugh, make merry, and enjoy our miserable existence on Earth. Solomon, in Ecclesiastes, talked about all his wealth, all his many accomplishments, and the daily life of man, and said everything is merely vanity and a striving after wind. That's where that section about to everything, there is a season, can be found. Great lyrics for that song…which band did that, the Byrds?

Anyway, a good example is the behavior of North Korea. Or a good example of absolute insanity…those guys don't exactly play nice with the rest of the world, and created their own special take on what is real, and what is decadent behavior that insults the blessed leader…when will those people wake up and realize their beautiful leader is screwing them left, right, and twice on Sundays, just so he can sit around and enjoy all the decadent western technology he tells everyone else is verboten, evil, and not to be lusted after. Selectively isolated from the world, their boisterous and disobedient ruler, little Kim IL Sung Junior, frequently pushes the envelope, instigating violent clashes with its Western neighbor, South Korea, just to see what he can get away with. His actions are those of a rebellious child. He is curious to see how strong his military actually are, and is willing to provoke a fight to find out. N. Korea's authoritarian big brother, China, has been repeatedly called upon by the rest of the world to keep him in line.

Exceedingly challenging and full of delight, with side order of deranged insanity.

While giving to others, you, and your entire being, feel joy and a unique sense of accomplishment. Emmanuel Kant said we are all born with the moral imperitive inside us, the ability to distinguish good from evil…so why doesn't N. Korea get with the program and join the rest of the world…it has to be a bit lonely, always marching the goose-step and carrying guns around, just so Kim can pretend he has a real army. Don't they all want I-pods, and I-phones, X-boxes, over-sized flat screen TV's, and Kentucky Fried Chicken? What's wrong with those people…all they think about is that 38th parallel…watching the economic progress of their neighbors to the south. They have to be jealous and feeling a little out of things. Perhaps they drugged the water supply, just so the brainwashing keeps working. Those parading tin soldiers remind me of the same parades Stalin had, and look what that madman did…anyone with a brain was transported, never to be seen again.

Question on new television show, Hanger One. Is the U.S. working with aliens and building a space fleet based on the technology the aliens give them? What is that all about…why are the aliens in charge…are they feeding us little bits of technology just to placate the army, and are really doing something that will benefit them. Are they trying to take us over with our help? A little tech here, a space ship there; they could be paying us off so they can manipulate us and achieve their own evil designs. Time will tell…sadly. Run for the hills…but that's where the aliens are hiding, so we won't have anywhere to run to, no place to go…dum de dum, sounds like a song.

Day 41:

Birthdays make you older, so keep them secret. You won't get presents, but, like Jack Benny, you can be 39 for a long time...wait a minute, he's dead.

Word play: playing with words...perhaps these would be excellent candidates for the new world dictionary...they've got style, and can mean just about anything you want them to mean. Look at what Lewis Carroll did for the English language...all basically, nonsense words, they are known by all: Jabberwocky, the Bandersnatch, Brillig, and the mimsy troves...did gyre and gambol in wabe. We now have the great hunting of the Snark...something that reminds me of Catcher in the Rye, with Holden Caulfield and his 'People-hunting cap,' always weird, as that's what that wacko that shot John Lennon had on him after shooting him. I always wonder what music he denied the rest of the world, as John had gone through a hiatus, and was just getting back into writing...when a genius gets back in the groove, you never know what might transpire. The world could be humming unknown tunes, like unknown tunes we missed from Jimmi Hendrix, Lynyrd Skynyrd, Buddy Holly, Jim Morrison, and Janis Joplin...the dead at 27 club.

Some words are just fun to say, and some don't follow...the meaning is arbitrary. Like sissified...nearest word is sister, or sisal, a West Indian fiber. Oh well, if the show fits...

I likee...Salsa. I like inspisate, to thicken by evaporation...waste of a good word with a lousy meaning—perhaps make your own. Sonorous: nice word, good meaning. Slattern—an untidy or slovenly woman. Can't use that everyday, you'll get slapped.

Then there's Latin, which always sounds "official." *Per incuriam*...danger in delay...or, better do it quick, or you're in a heap of shit. The modern explanation works.

We have to be optimistic, and thank them for the music they did leave us, and be happy with that, as you can't get mad at fate…it might get mad at you, and I'd personally rather stay under fate's radar…it has a bad habit of killing you.

Swift added Lilliputian, Brobdingnagian, the large and small people, and Houyhnhnms, the horse people…then again, you're not Swift, or Dodgson (Carroll).

It was a great parody on the traveler's tales genre, and Swift was a master at subtle sarcasm, John Gay wrote in 1726, in a letter to Jonathan Swift that "Gulliver's Travels is universally read, from the cabinet council to the nursery."

Some attempts at lexicography…maybe the start of my own dictionary…a collection of words that only I know, so I can talk, and only I know what I'm saying…that would drive a lot of people nuts, or done carefully, you can pretend to know a lot of big words...I know a few people that do this without realizing it, and do it on a regular basis.

Dance of the **seganchechers**, or whirl of the **dervishnuphychs**—nutty gyrations.

Equinoptechrial—an artificially created equinox, or an equal science…a science for ever occasion; an occasional science (hmm.)

Surreazzerany—a kingdom of surrealistic proportions.

Gnosticoptical or purezenegis—philosophy through a lens, or optical philosophical exterminator, and genetic purity from an egg.

Presbyteroquial or **Suggustinification**—a leader of equal standing, and implied importance, or implied effect from a certain object.

Then again, we have too many words already, and these aren't general water cooler remarks..

Always aware of the definition of insanity is good: repeatedly doing the same painful and negative actions that produces the same agonizing results, and never learning to stop. A frequent characteristic. Like Sisyphus's stone, roll it up the hill, and always rolling it back down the hill. A Sisyphean labor has no purpose and no end. Redundiniquious. A word that means…repeating yourself in a silly manner…a good enough meaning until I come up with something better.

Besides borrowing words from other languages, many words are adopted from literature, and therefore are created words that are merely put into use because they sound neat. Remember that short story you wrote, with the heavy acid user always talking with mixed up words he invented and only he really understood…since they were "portmanteau" words, taking half of this and half of that, they did make sense, and some seemed like they could be useful. (mumble…grumble and murmur…maybe not). What's this got to do with the price of tea in China…another saying that doesn't make much sense…the price of rice would sound better, at least it rhymes.

We all fall short of God's glory. He is the creator of all life; He is goodness personified, the Holy of Holy's. He cautions us that his wrath is terrible and just. We all know we deserve his anger, as we have not followed his laws and didn't worship him night and day, as we should.

I found that the argument for the existence of God got in the way of my spiritual growth. Claiming and believing that my sins were heinous and unforgivable, I was terrified of God's displeasure.

Then I remember the love God has for our world. He loved us so much He sent His only begotten son, so that whosoever believeth in Him should not perish but have eternal life.

I believe, thereby making all the other Bible laws and admonitions binding and applicable. I know I fall short, and fear the judgment I deserve. Then I remember I am a sinner, and that I must die to death and live for Christ. Only through the Grace of our Lord would I ever be accepted. His love has touched me, a lowly sinner, and if He can notice me in some small way, then His greatness is beyond measure, inexplicable are His ways to use, His mind a perfect thought that we could not comprehend.

Living for Christ and dying to sin makes me a reborn Christian, free of the stain of my former sins, and I should focus on praising his just judgments and sing His graceful wisdom. For who can fathom the endless mind of God, but God. We are but humble servants, who only want to praise Him and adore Him with overflowing love.

Day 42:

Well, you're over forty now, I think 50 should do it.

Wilderness insert: for the mystery of history story.

Alive with the sound of falling water, Topoff Lake echoed with the voice of nature, making the sound of a single gunshot ring with artificial thunder. The patter of rain continued, its constant tinkle faint over the roar of well-fed waterfalls coursing down the side of a snow-topped mountain. A large wood and brick lodge stood by the stream, its lakefront a transparent glass face presiding over the rain dimpled water.

A cedar and brick extension held a panoramic view of the lake and the waterfalls running alongside the house. A steady wind splattered the west walls with rain. The octagonal design was enclosed in glass, like a weather-proofed outdoor gazebo. A central fireplace, sloped as an indoor fire pit surrounded by comfortable furnishings, was Sarah's favorite place to curl up and read on a rainy day. Her love of the room would stay with her forever; she was now sprawled along the thickly padded divan, dead.

Quick switch to alternate location?

I heard that already," quipped Lance," get a grip on things. Life is a bitch and then you die. Now let's get a cab a get out of here." Slant nodded and followed Lance out of the bar. Around the bright lights, inky black alleys looked like dangerous conduits to another life. A bright red overtone was splashed with flashing yellow lights; every strip joint and scummy bar tried to outdo their neighbor with flashy neon promises. Names and proffered services turned Yonge Street into a long tacky uninterrupted light show.

Version Two:

A roaring fire and soothing blush of candlelight bathed the study's book lined walls in a gentle light, the flickering glow dancing against the well-polished mahogany and hardwood floors. Historical treasures held places of honor: a full suit of medieval armor stood in a corner, Victorian-era Persian rugs snuggled the hard wood floors while shelves displayed treasures from Egypt, India and China. There was something of interest wherever the eye roamed, giving the room an inviting ambiance, with cozy furniture to lounge on and enjoy the surroundings. Some of the museum quality artifacts had their own stands, with a mini-spot showcasing their beauty and rarity. It was a room that suggested secrets; along with the old and long forgotten mysteries, old manuscripts, books and documents that linked past mysteries with the present unknown. It was indicative of the room's owner, the Right Honorable Dr. Angus Darby, Royal Historian and Chair of Ancient Languages. Darby was a world-renowned expert on the history of Papyrus scrolls, parchment codexes and illuminated manuscripts, a document expert that had solved many riddles of the past, and whose work was often so confidential he had covert protection from Great Britain's top secret services, MI-5 and MI-6, the same services that protected the Queen of England.

PLOT COUPONS, McGuffin search.
A team killed her and searched the house for something.
She was questioned and killed, as she was related to Nina and her archaeologist Father.
They were part of the group that is seeking the solution to Voynich's manuscript.
Work in Darby, Sully, and the rest of Bernard's legacy; work out a timetable, geography, and times. Work in a denouement of suspense; plan a story of exciting revelations leading to the end.

It's not what you get, but what you give to God.

Not what you can do, but what you do for God.

Are you going to follow the word of God? If not, who do you think you are?

Are you too weak to overcome the sin in this world?

Must you hang on to sin?

He can do anything for you, so why not start now? Today, as Jesus said, do not go before the alter to serve God if you have something to correct with your friends. Rather fix it and come back to pledge to God what you must.

Once you get an artificial form of help, you are willing to go the long route. Why not stick to what God wants you to do?

What would you say to God about what He offered you, gave you and asked you to do?

Quit rebelling and holding to what was – for it is not. The world is not the answer, God is the answer, and you know the question.

Are you to weak to follow his way?

What is the matter with you?

God forgives by grace, what have you done for grace.

Are you waiting for the last minute, and keep your sights on this world? If you know His world and words, obey now, for it is what is in your heart that matters, not what you think is in your heart.

There is no such thing as being half saved; holding on to anything that is sinful in your mind is not holy, and only the perfect can stand in God's presence.

You die in your sin. No man can come to the Father but through Christ, and if you accept Jesus, you will follow his words.

You are saved. Go forth and sin no more, or something worse might happen to you.

Don't hold on to the world, or wait until pain and sorrow drives you to Him? You will always turn to Him, for that is your way and your belief, so why not start now and do something that might really help you?

You drink the water of this world for you felt parched, and again you are thirsty. Surrender all and you will have rivers of living water flow through you.

The path to Heaven is strewn with good intentions. Remember this, it has a lot more wisdom than words.

Day 43:

My muse is mumbling.

The Belly of the Beast

-1-

    For some reason I can't really recall, I started laughing. A real back slapping, gut churning, belly buster. I should have been crying, I think, but everything had a weird, surrealistic edge to it, like I was watching a real dream in a movie. I had no idea what happened. No problem, it doesn't matter; nothing matters in the aftermath of an ground-leveling, building-flattening explosion. Standing on the edge of this world and the spirit world gives you a unique sense of perspective. You suddenly realize how quickly things can end. At first, the sudden release of massive compressed power blows your mind; like ten thousand angry outbursts, you're never prepared to respond. Your eyes don't even have time to blink. The word stunned came to mind, and I began to ponder over what a suitable word it was. I was standing there, mouth and eyes agape, as shock waves radiated from the epicenter, now a drifting cloud of dirt, leaves and whatever else was in my house.

It was literally a bolt out of the blue; accordingly, I was completely dumbfounded, flabbergasted and well beyond the reach of any descriptive adjectives. Poof. I was there, walls around me; then kaboom, I was here, blown away, across the street, swaying back and forth after, after I painfully struggled to my feet. I looked down: my clothes were now disheveled rags, covered in blood, black soot and who knows what else. I was bruised and battered, but I don't think anything was broken. I had cuts and scratches everywhere. Along with a persistent ringing, everything sounded like I was in a long tunnel. I could make out loud yelling, which seemed to be coming from the crowd that was started to form around my former abode. All the neighbors that hated my late night parties and loud music were probably trying hard not to cheer. I must have looked like an explosion at the blood bank, and I think people were trying to ask me if I was okay. I just ignored them. I don't know if I was okay; I didn't know if I should be all right, or whether I should play up my injuries for all they were worth. My brain had been banged around so much I didn't know what happened, where I was, or who I was.

I just knew I had to come up with some explanation for the explosion, something that wouldn't get the cops sniffing around for chemical traces.

The whole thing had sounded like a good idea. Daniel always gets these great ideas; always incredibly dangerous, or blatantly illegal, he just focuses on his ultimate goal. Money, and lots of it. It always involves the rest of the gang, has Daniel safely on the sidelines calling the shots, and we stupidly do what he says. I would have said no a long time ago, but his schemes always seem to work, and we eventually get a lot more money than we had before. The only stupidity here is the way we always spend what we scam and never invest it properly. Well, it's sometimes hard to explain to the taxman how you suddenly got a couple hundred grand. And when Daniel explains his brainstorm, he always makes it sound like a walk in the park. They always seem to be so simple and straightforward we would be foolish not to try it out. Then again, everything sounds like a good idea – until it's too late.

Too late to put the cork back on that big, brown bottle Frank managed to sneak out of that chemical company. The one with the bright red label suggesting that its fumes and an open flame (the kind you light a cigarette with), might produce the cloud with all those lines jumping out of it. Highly flammable: I remember it well. Sort of an orangey-red label. Really stands out. Like random pieces of a picture puzzle, hazy images were filtering through my rattled neurons, and I slowly started to patch together what happened.

I'd been in the basement, mixing the chemicals we used to make the designer ecstasy we were illegally manufacturing, and I remembered using my trusty old Zippo to light a cigarette. I thought the fan had cleared most of the fumes, but it was a really stupid thing to do. Now I had to come up with two explanations. I couldn't tell the gang how stupid I was, not with Blacky freshly out on parole and counting on his share of the money to get settled. Blacky was notoriously easy to piss off, and big enough to do some serious damage before he realized you were a friend, not someone he could just kill. It was hard being friends with Blacky, but there was nobody I'd rather have on my side in a fight, or as a sidekick when trying some intimidating negotiation. A mean look from Blacky made people want to keep on his good side. Anyway, I could work that out later. Judging from the increasingly loud wail of sirens, my official explanation to the cops and fire department would come first. And it had better be a good one. Telling them I was cooking illegal drugs wasn't an option. I had to come up with a good reason why my house disintegrated; and I was pretty sure a gas explosion wouldn't cover everything. Maybe some extra BBQ propane tanks in the basement added to the gas explosion. Sounded like a good story. When my mental abilities partially returned, I got busy with text-messaging Daniel to tell him we out of material, and could possibly be in a lot of trouble if I didn't convince the cops this was an accident. I'd just tell Daniel everything was normal, and somehow a spark must have ignited the fumes. I also didn't want to leave a trail of cell phone calls to the gang. I sent Daniel exactly what he would expect if something like this happened. Dan. Gas…Boom. House gone.

-2-

The fire department were the first to arrive, a huge red hook and ladder job, freshly washed, polished and waxed. They had to do something beside sit around and cook. The new generation firemen and their bright-eyed, testosterone pumped enthusiasm was not what I wanted to deal with. They peppered me with questions, but I just pretended I couldn't hear them and walked around like I was in a daze. It was an easy act; my ears were still throbbing, and my pummeled body was sore and in want of rest and medical attention. I sat on the truck's bumper and stared off into space. I was going for the dazed and confused look. It would go along with the sudden explosion theory. I had taken a close look at the house before the first rescue team showed up. It was a catastrophe. There was nothing left but pulverized sawdust. These were the fire/disaster engineers. I didn't want to say anything to them before I got a chance to make sure it was what I should say if I was really in a gas explosion. Like how much did I smell, and if I had walked in and smelled gas, and why would Mr. Stupid immediately pull out a cigarette and flick an open flame? It sounded suspicious to me before I even said anything. I wanted to talk to Daniel first, but I had my story.

I had no idea how I survived. I must have been blow clear. It must have been the mattress I was standing behind when the fumes ignited. We used 7 mattresses standing end-on-end to make up the small lab's walls. They helped absorb any sound, or would help muffle any small lab "reactions." No one had any idea we could have the sort of explosive reaction I created. I must have been protected by one of the mattresses and lucky as hell to get blown out and away instead of trapped there and blown to pieces. I just hoped the destruction was enough to cover any chemical residue. We had a hydroponic greenhouse for tomatoes and herbs to help with a cover story, just in case something like this happened. At least we could explain large "carboy" glass jugs of chemicals as equipment for the greenhouse.

I thought it would double our risk; the greenhouse might make it look like we were trying to grow pot, but now I was glad I could use it as part of the explanation for the explosion. Anything leftover from the initial boom went up in flame and was just burning now. I gratefully noticed the fire guys didn't get the hose out; they were letting it burn out, as there wasn't much left to burn. That would help erase any residue. There was also a large plastic bottle that had over 3,000 finished pills. At 25 bucks a pill, it was a loss, but we'd already stashed over 80,000 pills and sold 60,000 for eleven bucks a pill.

The cops were the last ones to arrive; that was a good sign. I'm sure the neighbors would tell them all about the late parties, and mention the greenhouse where they were sure we were growing pot. Now some of Daniel's well thought out planning made a lot of sense. The cops would know that if we had something to hide, the last thing we would do is have a huge party and attract attention. They'd also know most grow-ops were in basements, in houses that didn't want attention. The idea was that the cops would think it was a neighborhood domestic problem and not something that was worth looking into. Doing the exact opposite of what other drug makers did was our plan; we'd find out pretty soon how well it worked. Some of the neighbors had made friends with me, and I'd apologized for the parties, so they might get conflicting reports. That would work out well. I know some of my neighbors were busy-bodies that spied on everyone on the block, so they'd no doubt take this chance and tell the cops that everyone on the block were selling and making drugs. Some would be terrorists making bombs, and others were probably high-class criminals, doing diamond heists. The cops would shake their heads, thank them for the promising leads and get the heck out of there as soon as they could. I thought it was a good cover. Nancy and Tom showed up, acting the concerned friends, but tried to get me aside and ask what really happened. I think they were more worried about the product than whether I had limbs blown off.

A young, fresh-faced officer came over and asked me a few questions. I cupped my ear, shook my head a few times, and generally tried to look like a complete victim of home gas explosion. I must have done well, as he left me alone and went about his business. Without staring, I followed his progress around the scene, along with the cute red head that was his partner. They both went over to the house, had a look at the extent of the destruction and went back to talk to the firemen. They were the experts here. That was a good sign. The last thing I wanted was for some CSI freak to start picking up pieces of glass, smelling them and then bagging them for chemical analysis.

I'd occasionally been accused of being a technology-hating Luddite, but it wasn't the technology I hated, it was the machine like efficiency I saw rubbing off on the people that used the technology. I really missed the laid-back attitude of the '70s and 80's. Even the '90's were tolerable, but everything in the new millennia was centered on a silicone heart and a digital response. If this was the future, I didn't like where it was heading. People were more friendly and easy going before everyone started using text messaging, e-mail, and cell-phones for social interaction. People forgot how to talk with each other and were uptight, performance focused, and goal oriented. Money and technology; technology and money – they went hand in hand.

Finally, the medical response team arrived. I recalled there was some strike action going on between their union and the city. Time to put on my innocent-as-can-be game face. I decided the best story was the one we had all agreed on and had even rehearsed. I'd already told the firemen I had been in the kitchen, lit a cigarette, and ended up on the other side of the street, adding it must have been a gas leak. A tall blond haired man with a incongruous black mustache put his red medical bag on the ground.
"How are you," he asked eying me up and down "I'm Tom with the paramedic service. Do you have any major injuries?"
I put my hand behind my ear to indicate my hearing problem. "Did you ask me how I was? Sore. All over…and I have a ringing in my ears. I'm badly shaken."
I leaned towards him to hear him.

"That I can understand. You survived that?" he asked, gesturing towards the house.

"Yes sir. I think I was blown clear."

"Anything broken or have any internal injuries?"

"Broken? I'm not sure. My ribs are sore, but I can move everything."

His partner, a small Asian woman with a tight ponytail rolled up the ambulance's gurney.

Her eyes widened when she saw me. I remembered I was covered in blood from dozens of cuts and scrapes.

"Hi there," she blurted out as she popped the gurney up. "My name is Nancy. We want to take you to the hospital and have a doctor look at you."

"Thanks, that sounds like a good idea."

The paramedics got on either side of me and helped me up on the gurney. Nancy unfolded a red blanket and tucked it carefully around me, adjusting my pillow and carefully strapping the seat belt around me.

"Okay, here we go."

They rolled the gurney over to the ambulance. Before they tucked me inside, I had a look at the house. The firemen were still watching from a distance. The police were talking to the crowd that had gathered. Nancy climbed in beside me while the guy shut the doors. I think Tom went to talk with the cops. Nancy pulled out a notebook and started asking me for my health information. I offered her my wallet; she took out my health card and driver's license. I closed my eyes and relaxed. I heard the driver's door close. He started the engine and we were under way. So far so good, I thought.

Day 44:
Rambling, gambling and risking all.

Children start loving their parents, as they grow older, they learn to judge them: this is the fork in the road, when you forgive them for the difficulties raising you, and feel you're special, and have that fatal urge to judge people…no one should judge, as everyone has flaws, and only hypocrites think they'll good enough to point out other's errors.

True love is patient and kind; love doesn't envy or boast; it is not arrogant or rude. It does not insist on its own way; it does not rejoice at wrongdoing, but rejoices with the truth. Love bears all things, believes all things, hopes all things, and endures all things. For that sums up the new law: love your neighbor as yourself, for no damage comes from unselfish love. By the grace of God, I am what I am, but I know I will do more for the Lord. Do not be overcome by evil, but overcome evil with good.

If possible, so far as it depends on you, live peacefully with all. Never avenge yourself, for God said, Vengeance in mine, I will repay. It is a terrible thing to land in the hands of a living God, after rebuking him your whole life. It would have been better if you never heard of God…if you have, no one will boast, brag, or argue in front of God.

Memory is hard thing to forget, and harder to forgive; some refuse to forget, and recall their own version of events…sometimes these people are incorrigible, unable to change. Yet, despite the horrible, abuse you think this person has done to ruin your life, there might come a point when you have an epiphany – that you shouldn't JUDGE, lest ye be judged. Never be conceited, for we all fall short of the Glory of God. Remember, if you refuse to forgive someone, how is the Lord going to forgive you? After you have forgiven and forgotten your burden: you'll find life a joy to live, more delightful each day, and your personal anomie is poignantly altered. Honesty, especially with yourself, takes a big man to

The strife of brothers can be brutal and cruel – they seem to be on fate's great balancing beam, where a microgram here, or a milligram there, upsets the whole equilibrium. Too bad, it couldn't just be laughed off, as life is too short for bitterness and resentment. I've has friends that we both left, saying we'd never want to see them again. We bumped into each other four years later, and all was forgotten. Actually, I don't remember why we were fighting…it was that stupid.

When King David gave us his words of praise to our Lord, he definitely knew the problems that can arise between Brothers.
"Behold," he wrote, "how good and pleasant it is when Brothers dwell in unity, like the precious oil running down the beard of Aaron." Brothers seem to be another one of those 50/50 splits – you either love him or hate him.
What are you waiting for? Surrender your whole self to God, and he will help you fight your battles, in fact there are no problems God cannot help with. But, you have to surrender, get down on you knees, and say "God I want you in my life, I need you to help me fight these unwinnable fights I face every day." You need to ask Him into your life, totally, willing to do anything for His will, through yourself on the floor and pray that He will accept you, a terrible sinner."
Run, don't walk, and ask for forgiveness with a true and open heart, surrender all and receive all, only then will you understand the Power of our Great God.

If we could know the glory and path that God had planned for us, we should not be so foolish, and run to the alter, surrendering your will totally, without reservation, so He has a chance to do something in our life.

What is unconfessed? What is holding you back from the Holy Spirit lifting you to new heights?

Everyday, we should be on our knees, surrendering to God, thanking him for our gifts, and praise Him for being such a loving Father.

You need the surrender completely, every iota of your soul needs His love; with this surrender, we can call on him to help, to open doors, seek and to find, ask and it will be given. If what we need is an outlet for love, call on Him, for He will help, but only if we give unconditionally, freely, openly, to obey his Laws, walk in Jesus' steps, live a life of love. Change now, and enjoy the gifts the Father wishes He could give you, but He will only give them to those who accept Him entirely, without any doubt.

Surrender your body, give up your soul, turn your mind, focus your emotions, and He will make a difference in your life.

Pray on your knees, with total commitment. What are you waiting for? If you have knowledge that the Lord is Good, run after that goodness, and depart from sin.

Become a slave to righteousness, not to sin; only then can the Father work miracles, for you will become a Child of God – and what Father, if his son asks for bread, would give him a serpent?

Do not wait…look in your heart, find what is still a problem, and ask the Lord for help. He will help with all battles, like drugs or smoking, emotional problems, lack of love, or understanding. He will give all, if you surrender all.

What are you waiting for? Enjoy your life while God allows you to, and enjoy it the way He envisions for you, for it is more glorious than anything I could ever think of.

Listen to His words, understand, and pray for enlightenment. Pray for your soul. Accept Him today, WHAT ARE YOU WAITING FOR?

Surrender yourself; give yourself as an offering for God.

DO NOT love pleasure more than God. What does drugs do to you? They have lost their magic and only give you pain…why stick with them. Ask the Lord, and He will let you know, deep down, what you need to change to be acceptable to Him.

Don't focus on this world, focus on the spiritual; what good is it to conquer this world, when, for all eternity, you loose your soul.

Don't focus on money, focus on God; he will provide, all you need will come from the Father, don't expect great riches for becoming His child, be happy only that you are one of His children, a gift that is priceless.

In Jesus name I pray,

Amen. That's the way I was taught to pray, and I think it's the best.

We seem to be focused on conquering the unheard of, exploring the unknown. What about conquering greed, eliminating prejudice, embracing equality, striving to understand humility, purge false pride and replace selfishness with charity. Our world has always needed something to believe in, a reason to live, and a challenge to stimulate us. We already have a major challenge. Eradicate sin, help each other and worship God zealously. We are encouraged to outdo one another in showing honor, and lead others to the word of God with zeal.

Basic plot outline:

A street drug becomes a special formula to change life. It starts out as some new ecstasy/opiate derivative, and the formula was a hit-or-miss hodge-podge of chemicals—the chemist lost his notes in a fire for the exact recipe needed to make all the quantities needed for some of the drugs to interact. It was made by a brilliant doctor who was a pharmacology graduate. He wanted to make a drug that would fill in the gap to make people feel better, but not be addicted to the substance.

After experimenting on himself during an injury, he combined high levels of vitamins with strong painkillers and was healed quickly and felt better after the accident than before it. He knew he was on to something, but realized experimenting with what he took would be a problem: clinical diamorphine and oxycodone would present many problems, because of the stigma attached to heroin, along with the difficulty of getting it. He experimented with rats and rabbits for a year, finding a certain mix of powerful opiates (fentenyl), along with vitamins, and other drugs that promoted good health, happiness, and mood stability. He envisioned a wonder drug, using unconventional and disliked substances, combined with health promoting drugs that were proven and accepted in the medical community. It prolongs life, and improves life.

Possible scenario:

He ended up working for a major pharmaceutical company that supplied almost ever drug in the Blue book. He had access to pure pharmaceutical diamorphine (heroin), cocaine, muscle-relaxants, tranquillizers, mood stabilizers, antidepressants, and other drugs that helped release neurotransmitters that aided the body in producing natural endorphins. He also used some street level heroin that was 70% pure, mainly because a fellow colleague, who was also a heroin addict, found a source of brown heroin from Egypt that was made by an ex-pharmacist and used clinical grade drugs to turn a special batch of top-drawer opium into heroin. There were two types he had in large quantities: the pure heroin, and an experimental mix that included some of the ingredients of MDA.

He was able to get two kilos of each.

He mixed them all together after doing a test batch that worked well. When he made the final mix, he also added aspirin and binders, and even threw in pure vitamin B, D, C, and a b-12 tri-ox compound.

Starting with the 4 kilos of illegally obtained heroin and heroin/MDA, he added almost equal portions of the clinical drugs he "borrowed" and basically used the factory at night, when it was closed down. He made a huge batch of it, using binders and clinical fillers to help balance out the dosage and help turn it into pills. There were just over 20 million pills made, each containing about 100 milligrams of his special mix. These were the pills that boosted the enjoyment of life, and would eventually make people live an extra 50 years.

The drug cures all diseases and makes someone perfectly healthy. The drug is eventually found to be the cause, but by then, many people just used it to get stoned, and there was no major supply. Because it was illegal, dealers were reluctant to admit they had any. Finally, it was linked to these miraculous cures turning up all over the country: only 3 people with any quantity, and a few that bought enough personal supply to last them for years.

The drugs made someone feel great, enhanced abilities, boosted IQ levels, retarded aging (discovered after 6 years), the long term effects were projected. The proportion of heroin and cocaine was just enough to stimulate the brain's pleasure centers, but mixed with permanent anti-depression drugs and the mix of vitamins, tranquillizers, muscle relaxants and such, the areas of the brain were stimulated in a permanent manner; the endorphin production area along with other areas of the brain were affected for good.

Doctors and politicians slowly found out what was going on after years of inexplicable discoveries. Down and out heroin addicts, homeless, single mothers, welfare recipients, and occasional drug takers were the only people who showed signs of a new virility. Of course, the rich and famous were terribly jealous, demanding to know why these "low-lifes" were enjoying something they didn't have.

A take on the old "payback's a bitch" attitude.

It was dismissed for years as a weird by product, but when the data on thousands of patients was amassed, after 7 years, they figured out a mutation had occurred. When they tried to isolate what could have caused it, this ecstasy drug was the only thing they had in common.

Most of them only used it once or twice, and then many of them gave up drugs, as they were in no need of the thrill of a high, as they were already feeling great.

Perhaps, an added twist.

The drug was the invention of Tim O'Leary, a Doctor who specialized in pharmacology; for years he had been trying to create a "happy pill" that had all the benefits of opium with the energizing effects of cocaine, tempered with the calming effects of tranquillers, anti-depressants and mood stabilizers. It was medically beneficial with the addition of vitamins, oriental medicine, muscle relaxants and a host of other generally helpful drugs. He believed a wild cocktail of progressive drugs could stimulate the brain into releasing natural endorphins and anti-toxins, along with other naturally produced hormones that would improve general health and well being. He worked successfully with lab rats and other experimental animals but was stymied in his research by large pharmaceutical firms and the constraints of the Food and Drug Administration.

The problem was with his choice of medicines. Including heroin and cocaine into his cure-all drug incurred prejudice and disapproval by the draconian laws against illegal drugs.

O'Leary was a visionary. Looking beyond the stigmas associated with the use of cocaine and heroin, he concentrated on the undeniable effects the narcotics had on certain areas of the brain, and wondered what benefits that effect could have if used in moderation with other drugs that would combine to provide an unknown interaction.

The mix of mood stabilizers, anti-depressants, narcotics, stimulants, vitamins, and other drugs with positive effects on our human chemistry, he created a pill that was more enjoyable than the fabled fountain of use. A Senator got wind of the whole affair and asked why this wonder drug was not available to the general public.

When told it contained heroin, he replied "I'd take arsenic if it was proven to do everything this drug can do."

How long before everyone wanted it, no matter what it contained, and no matter how it conflicted with religious beliefs, like the Luddites that spurn modern medicine and won't even get blood transfusions…preferring to pray and trust in God.

Trusting in God is good, but giving something for God to work with is better. (see short story: "The Cure of Cain").

What would the world say to a wonder drug like this, especially if they didn't know what was in it, and only knew it had major life promoting qualities, like the fake cure-alls they sold off wagons in the old wild-west.

Just an idea.

To focus our attention on technological advancement and worship at the altar of science is sheer arrogance; we forget to worship our creator. Don't hide from the world; embrace it and marvel at the flawless perfection of creation, astonish yourself with its palette of scintillating color and overwhelming glory. Enjoy the beauty that is alive with splendor and grace. Alive with a magnificent menagerie that are intricately bizarre and incredibly adapted to accept an unknown function that somehow contributes to this mountainous maze of interlocking grandeur and grace.

I found out the hard way doing the right thing doesn't necessarily make the thing right. I'm all for law and order, but I also believe a little common sense and ingenuity should be able to jump the queue in some cases, especially when the good of all humanity is at stake. The only thing I've ever seen pushed to the front of the line is greed. When it means lot of money for everyone involved, safety and health approval testing can be expedited, usually sacrificing something in the process.

There was money to be made in what I wanted to do, but I didn't want to make money; I wanted to give something to humanity. It is, after all, better to give than receive. Ironically, that wasn't the only biblical lesson that came from this. The poor-in-heart, the down-and-out and the meek were the first to benefit from my discovery. It took the high and mighty some time to realize that rules were guidelines, and shouldn't teach us how to think. Our minds are able to do that. When rules and laws affect our attitudes, we become nothing more than prejudiced machines. What is right and what is wrong should always be determined as a matter of morals, a decision of the heart.

It's been said that bureaucracy is there to protect us. It's responsible for evaluating and examining new products, making sure they are safe for public use or consumption. The only problem is it can take years to just make the decision that a decision should be made. It is a slow and ungainly beast; protecting us with multiple layers of security becomes a lengthy procedure and sometime does nothing but provide an entire department to blame if something goes wrong, instead of just one person.
Why do bureaucrats so keen on shifting the blame…they're not elected, but they certainly make sure they can shift the blame whenever they want. No one has the balls to stand up, admit they were stupid, sit down, and just apologize. How hard is that.

Day 45:

Fortune favors the bold, so we are told; redemption favors the good, so I've heard.

The rise and fall of the 20th century:

Doomsday. Adios Earth…goodbye cosmos. Our scientists want to understand the secrets of the universe, and by experimenting with forces they don't understand, they could cause the end of everything…thanks to curiosity. Smashing atoms together can create forces beyond our physical comprehension of how things work. When atoms collide, as in the Large Haldron Collider (LHC) in Geneva, can form particles that some might say should remain undiscovered until we have the scientific ability to control them. Strangelets, or microscopic black holes, can form in their quest to create the Higgs-Boson particle, arrogantly named the God particle.

When playing with fire, man gets burnt; if playing with the underlying fundamental structure of the universe, man could cause his own destruction. A singularity like a black hole can grow…eventually growing so large it could consume our planet. They also form in extra-dimension we do not understand, but only know about from some elegant theories. Eleven dimensions are believed to exist, and our Creator resides in one of them. Playing around with his private dimension, his personal "rest," has many nasty implications. We don't want to anger God…by creating us from dust; he could very easily return us to dust.

Oh the hubris of our intellectual nutbars…they are playing with things they have no right to…opening boxes that Pandora learn best remain closed. Humanity should first understand our past, for through learning what has been will give us a better idea of how to deal with what might be. There are so many mysteries in history, we would be morons if we didn't learn from them, and find out the truth behind our existence. Eternity is a long time, yet billions of years are but a hiccup along the line of forever, and there are many unknowns archeologist ignore, cover up, and refuse to accept the implications of what they mean. The evidence of past extraterrestrial intervention in our evolution is scattered throughout every written history on earth, yet it is ignored and dismissed as irrational and foolish. By ignoring these signposts along our living history, we are the fools, and we will reap the benefits of fools unless we take a quantum leap forward and imagine the unimaginable. We can't even deal with our foolish human emotions.

The writing has been on the wall for a long time, but we consistently mistranslate, ignore, and fail to understand the ultimate message. Despite years of speculation and study, the ultimate fate of mankind will be determined by good old human nature. We still suffer from humanity's greatest enemy: itself. Greed, corruption, and complete lack of morality has always been our Achilles Heel, and our destruction will hinge on the same evil depicted in the Bible, and the technological advances made since the time of Christ will only complicate our end, and not save anything. The fate of our world depends on numerous scenarios, religious fatalism, philosophical differences, and the willingness for the individual to sacrifice his fate for the overall good of the community, something soldiers experience in battle, and have learned is not an easy expectation. That's why we call such people heroes. Falling on a grenade to save your buddies is an instinctive act, and one that requires someone to accept the fact that they are a spiritual being, and that spirituality will be our salvation or our destruction. Uniting the world would take a miracle, an event that is beyond something we can understand…and that could very well be the apocalypse we are told about in the Bible. This sort of wisdom belongs to God alone, and with so many faithful followers that believe in different Gods, we are faced with a situation that has no solution.

When examined in that light, our future can never be functional, and a war of all against all will be waiting for us, just as sure as the sky is blue…for now. That too may change…in an instant. Philosophers have suggested that events like an attack by aliens from another world would unite humanity, but that still does not change the problem with the true nature of man. Any invading aliens could get help from a group of humans just by promising them safety, security, or something they want. Never underestimate the greed of humans, their innate selfishness, and their inherent greed…people like this would sell out other humans in an instant. When you look at the survival of the human race in this manner, there is not a lot of hope, as even hope is personal, and staying alive to hope another day might require murder, deception, and falsehood…all of the old evils pointed out in the Bible, and since these points are still alive and well, and their existence could be the reason for the end. The only thing that makes us an odds on favorite for survival would be selecting a group of humans that obey their moral code, and would be willing to give that last drink of water to their children, so they might survive. Unless we're willing to die so that others may live, I wouldn't bet a lot on the survival of humanity in any situation that puts survival in a Darwinian sense…only the strong will survive, and the strong are willing to kill to ensure they have the last sip of water. So much for our great society…we will kill ourselves in the long run, unless God is trusted and obeyed, for He is the real determiner in all cases. But, we even have atheists that don't believe in God, so their morals wouldn't save anyone except themselves. With observations like these, it's no wonder so many civilizations have risen and fallen, but since we are now a globally connected planet, the real question is whether another, out-of-this-world disaster, would leave behind tiny pockets of survivors, or whether humanity would succumb to the Biblical end that is prophesized in Revelations. Another major fork in the road is the predictions we read about in the Bible. If we trust that it is the word of God, there is nothing we can do to change our ultimate end, for God doesn't lie, and Revelation states that this world will end, to be replaced by a group of morally correct humans that all believe in God, and their righteousness will be their salvation. Facts like these are not open to debate…the scientist that thinks he can cheat God are in for a rude awakening.

History is a great teacher, and shows us that every great civilization has enjoyed boom times, and were eventually destroyed by the very expansion that made them so great. The adage "Those who do not pay attention to history are destined to repeat it," still contains the existential warning it always had, and human hubris or belief in their technological achievements to save them contains the same seeds of destruction that erased humanity's greatest civilizations. The Mayan, Roman, and Anastasia, plus unknown societies that have been eradicated by time, are all victims of similar thinking…trusting in their mastery over nature, yet ignoring events that finally destroyed them. The Mayan cut down too many trees, experienced agricultural failure, and grew too large to sustain them. Roman society fell over similar reasons, a failing agricultural infrastructure, but ignored the threat of marauding bands of barbarians that were mobile, vicious, and lived off the land by stealing everything they needed. Perhaps that fate may befall 20th century man, for it is cheaper to stay mobile, live by victimizing static and productive societies, and stay ahead of the game by becoming militarily stronger.

Our philosophers, theologians, and anthropologists claim that humanity's moral code, to live and let live, will go out the window as soon as we no longer have the necessities of life. Rational thought includes stealing the means to live, especially when a certain class or part of society holds onto these resources. We had two major world wars in the last 100 years…what will happen when the very essence of life, food and water, are hoarded by a few, and denied to the multitudes. Morality will be discarded, and the fight to live will replace everything we have supposedly achieved in the last 2,000 years.

The destruction of Rome is a giant signpost, warning future societies what will happen unless equal division of all resources is enjoyed by all. Our modern society has become so advanced, common sense is discarded, and we trust in our laws, police, and armies to ensure law and order will survive.

There are so many things wrong with this viewpoint, it's a wonder that more lawmakers don't realize this reality, and thanks to their stupidity, we will deserve to fail, just as so many pitiful human efforts have failed in the past. Ignoring the Wisdom of Solomon, messages written in boldface by great thinkers of the past, are neon signs we are too confident to understand, or too preoccupied by our success to comprehend. When Solomon wrote "There is nothing new under the sun," he underscores the lessons of history, past facts that have no place in our modern, face-paced world. Unfortunately, unknown realists that know what can instantly end the reign of proud but blind leaders that only focus on maintaining an unsustainable lifestyle, and will be the downfall of everyone.

When nay Sayers, scientists, and people that can predict the future are ignored, we are stuck with the often self-serving leadership of morons that can't see beyond their nose, yet they are in control of all our resources. Some of these realization seem so obvious it's a waste of time to write about them, but unless enough people see what's coming, nothing will be done about anything. Even if this becomes a forgotten piece of archeological discovery, uncovered by a future society that tried to understand why a globally aware population was unable to change and stop a head on disaster, it will have served a purpose.

During the cold war, Russia and the U.S.A. accumulated enough nuclear weapons to destroy Earth over and over again. The "Mutually Assured Destruction Pact," or MAD, gave our leaders the ability to rain destruction upon every living soul on this Earth, and that was so unacceptable and unreasonable, even our blind leaders saw the insanity and endeavored to reduce the nuclear stockpile. We still have enough to end the world, but hopefully cooler heads will prevail, and try to save humanity from annihilation. The people called dreamers, or "preppers," are groups of people that are trying to save themselves with limited resources, but seem to have a selfish survival goal, and do not have the overall continuation of the human race at heart. Stockpiling guns, preparing to defend their limited resources after a disaster, is the same sort of thinking that will doom us all, and sadly, it seems that they might be our only hope for survival.

Problems always have solutions, but saving humanity may not be a problem with a solution. Surviving an initial Extinction Level Event seems to be a solution that will only prolong our eventually demise. The world needs to put aside petty differences and begin to live as a thriving planet…that miracle is a dream, thanks to human greed, selfishness, and pride. We have not matured enough as a race to survive. Certain political situations, religious differences, and the overall division of priceless resources are all globally coveted realities that bring out the true nature of man.

After the warning of 12/12/2012, many people recognize that a quickly unfolding chain of events could turn our once beautiful planet into a morgue. Predictions have come and gone, and have taught us a valuable lesson, something right out of the Bible: no one knows the day of destruction, except for God. Our civilization could exist for another hundred years, one day, or thousands. We will never know. If we are to turn to our source of morality, the Bible, we shouldn't get all worked up about our eventual fate, for there is absolutely nothing we could ever do to change God's mind, unless the whole world embraced His word, repented their sins, and lived to worship and honor

God, always willing to lose their life in order to gain it back through true fate. The odds on that happening are so astronomical, any betting man wouldn't even hazard a dollar. The human races has too many problems to be saved…therefore, our fate is assured, and we should learn to accept it and live life with the best intentions, becoming humble and servile, pleasing to God, and as morally correct as we know we should be. Conquering our faltering morals sounds like the best way to survive, for with humanity, many things are impossible, but with God in control, there is nothing beyond what He has in store for us.

It seems the bottom line isn't building a space craft that could be an ark of sorts, but learning to fix the moral iniquities we face, and live the way we should live. Because there are so many people that faithfully believe there are more than one correct way to live, there is no way we will all survive, and only those pleasing to almighty God will live to see another day. We can argue until the cows come home, and philosophize until the end of time itself, but knowing the fate of the human race is presumptuous at best, and impossible in reality.

Day 46:
The days are getting shorter, my eyesight now grows dim.

Post-Structuralist Theory: from chaos to order.

Notwithstanding endless second thoughts, waffling positions on issues, or internal turmoil within our minds, humans like a neat and orderly environment. People like a methodical world…we plan and clean our cities, our apartments, and whatever we call home. A lot of people accumulate stuff, and if they want to have instant access, they store everything in a neat and logical manner. Imposing order on their chaotic accumulation of unique and unrelated items gives them a sense of control, a feeling that they are in charge of their lives. Occasionally, what we treasure explains what we value in life; one man's junk is another man's treasure, a truism that never changes. Certain people can look inside, and recognize the eternal values of correct morals, counting material possession as junk, avoiding the pitfall that things begin to own you when you over-emphasize their importance in life. Toys and playthings are worldly goods that are meaningless; we live on this earth to learn about life, not accumulate massive collections. There are a few neatniks that alphabetize record collections, household expenses and records, and even store their tools on peg-boards with the shape of the tool outlined in marker. This behavior is similar to humanity's overall need to have a well-structured life, understanding the components of that life, and grasping the basic rules that explain how our world functions.

Like the Structuralist title to this essay, humans like to impose order on what would be a chaotic mess without the science behind our world…something great thinkers do when they try and understand the concrete and meta-physical world around them. Invisible forces, such as magnetism and gravity, exist in the meta-physical world, just as human thoughts belong in dream-like environment of the mind. If we could see the invisible forces around us, the world would look entirely different. Plato, one of our world's earliest philosophers, proposed that everything with substance exists in a world of forms, and by tapping into this world, we are only extending a meta-physical concept by translating it into a solid entity. For Plato, the essential thought or idea of a wheel exists as a impalpable form, and when making a wheel, we are only borrowing a universal idea and turning it into reality. Accordingly, these thoughts exist in an ordered sense, where one thought leads to another, and so forth. Turning these thoughts into a document is a similar process: writing elevated the world's progress by recording thoughts, refining these thoughts, and enabling them to be passed on, so other people can add new thoughts to the collective wisdom of the world. The invention of writing is one of the basic elements that allowed mankind to progress from living in caves and make-shift huts to building cities and establishing an organized society that trades what is has for what it doesn't have. Trade is the basis for mercantilism, the economic engine that allowed Britain to help shape our modern world. That skips a lot of history, timely events and inventions that slowly built our world, but this is not a step-by-step historical record, it's a commentary on the human condition. Whether this is a manuscript that improves our shared knowledge always depends on what a person already understands…this is not an essay on cutting-edge modern life, but an opinion on the way our world works, presented in a logical and easy to comprehend manner.

Philosophically, structure implies a human projection of order on the often chaotic nature of the world. This order relates to the micro and macro world around us, systems within systems, purpose applied to environments that are beyond man's ability to understand. Therefore, imposing order on these systems allows people to think they understand what is going on around them. The microscopic world teems with life — life that we can observe, categorize, and therefore apply meaning to the entire system. Scientist describe self-sufficient eco-systems that are innately needed to fuel a larger system, often in ways that are mere guesses, but are accordingly systematized, ordered, and given meaning.

Some of the more bizarre meta-physical interpretations of our world state that there is no meaning to these functioning worlds, as meaning is a human projection, usually applied to complex phenomena to help our human understanding of something beyond empirical science to explain. We don't know why there are hundreds of sub-species of beetles, other than the reasons we attach to their existence. Beetles are a wonderfully designed and complex insect that are integral parts of the natural world, and studying this world adds to our overall understanding of the universe, the many complex systems that sustain our world, and hopefully discover something that will ultimately be useful to mankind. The natural world, with its complex fecundity of chemical interactions, could be the source of miracle drugs, helpful creations, and allow us to comprehend some of our planet's intricate functions. The world of bugs is all about released pheromones, pungent chemicals, and even organic electrical field created by living organisms and their evolutionary systems of self-protection, mating, or methods of getting food. Another world that would look amazing if we could see the intricate, invisible sub-systems in which they thrive.

Science has shown us systems within systems within systems, in an almost endless chain of life that feeds upon specific elements to maintain their existence. Man's need to impose meaning and order on these systems allows science to discover new features within these functional apparitions, and these discoveries are analyzed for anything that might help the human race. Plants have their own complex systems, discoveries that allow scientists to further their understanding of chemical interactions, as in the case of plant pherhemones, chemical reactions that sometimes protect the plant, in the case of strychnine, or certain markers that let trees thrive in complex systems that include bacteria and insects as natural fertilizers, propagating tools, and nutrients. Those important aspects of this natural system might hold secret benefits for man, in the form of new medicines and biotechnology. Understanding this microcosmic interaction would give science a valuable step forward in discovering the mysterious interactions of these systems on the larger system of which they are only a small component.

Philosophers constantly debate the meaning of life, its many sub-categories, and hope to expound upon human existence and mental capabilities. Within the many sub-sections of philosophy, epistemology is the section that deals with knowledge, analyzing how man acquires these facts, and how his mind interprets their meaning. When philosophers begin slashing all awareness to shreds, taking an existential view on life, they are merely debating how we rationalize data, and try to explain how our brains are able to perceive anything that exists outside of our personal thoughts.

Sometimes I thing the old "Occam's Razor" argument should be used: do not complicate any theory beyond necessity. Or, keep it simple stupid: when you have a simple concept to remember, your brain is free to associate with unlikely elements, and therefore conclude something that has never been realized before. Like the discover of microwaves: someone left a donut by some electrical device, and noticed a build up of heat. Then again, some people think microwaves were alien technology we got from wrecked spacecraft. I think the donut explanation seems more likely, but if I was writing, I'd prefer the ET explanation. When you're trying to entertain, always go with the myth of legend—the facts might be the real story, but they're boring and don't make exciting tales.
Another scientific lesson is:
If first you don't succeed, try, try, try, and try again. Practice makes perfect.

Day 47:

Another one bites the dust.

I think astronomers believe in God moreso than chemists: chemists witness predictable results, whereas astronomers see the infinity of His creation, and know there's a lot of stuff out there we do not understand. Philosophers just think whatever they want. The rest of us either believe, don't care, or are so boffed on their drug of choice they are too stoned to make any decisions.

What is the attraction that holds matter together? Is it part of the same force that holds the moon in orbit? Good old gravity? Does any mass, no matter how small, create a force that attracts and forces objects to endlessly circle them? What started this motion? Something like the big bang theory that we say started our universe?

Was it the hand or spark of God, touching a super dense piece of matter that ultimately spread out into our universe? Only God knows these answers, such as where that super dense chunk of every element that creates life come from…how can a scientist ever understand that answer? He may guess about the process…the chain of events that occurred after that initial spark, but showing where that matter came from is beyond science, and in the hands of God.

With such a sure example of the Divine handiwork, why are we not on our knees all the time, praising God, helping others, and doing what He suggests we do with the life He gave us?

Does evil interfere with our free choice? Or, are we unable to resist temptation, fall into sin, which is merely a part of life we must live with, after we learned about it in the Garden of Eden?

These questions, and others, will never be answered by anyone on Earth…the only person we can trust in this is Jesus, who left and returned to Heaven, bringing us the Word of the Father.

Will the world end on December 21, 2012? God said the end is to be better than the beginning, so an end is inevitable, talked about in Revelation, and talked about by Jesus, our Saviour, and high priest of humanity. God does not lie, and would never talk about such an end to civilization unless it was a fact. When God says something is a fact, we can take that to the bank. What should we do? Acceptance of our fate is in accord with what our creator has in mind, and anything going against what our creator wishes is going to fail. He made us, therefore He can do what He wills with us. The mold does not ask the mold maker to change what His plans are, that is entirely up to Him.

Building a super-spacecraft is almost a disaster waiting to happen. Unless God allows something to succeed, it will fail, and fail miserably.

There is no way to cheat fate. We consider our freedom to be ours, that there is an alternative life to the one God tells us about in the Bible, but if the Bible is the word and plan of our creator, we can only submit to His will, and trust in His judgment. This is not open to debate. Does this mean we should embrace our death? God did say the end should exceed the start, so thinking along this line is a good idea. We cannot escape death, and only live with what we are given.

We should pray, ask for forgiveness, and turn from sin. That, in the long run, is the only way we are able to ensure eternal life. Trust in God, and do exactly what He says what we need to do, and that is erase sin from our lives, for He is Holy, therefore we must be Holy, and try to achieve this as our main goal in life. Simply put, it is the only goal in life, and we must fight Satan, accept goodness, and say NO to sin and what might go along with that.

Living with your main goal driving you; do not question it, accept it, and trust that it is for the best. Thinking we are missing something is vanity, pride, and sin. Be humble, be good, and worship the Lord. He knows what is the best for us, and we must embrace that with open arms, and a glad heart. A free heart is an open heart, and we must understand our place in the world...be willing, humble, and full of the gift from God. That is the best thing we could achieve, and the only thing that really matters. Accept this, and you will learn to love life, lose fear, and embrace whatever God has planned for you. Trusting in God is beyond our understanding, but it is the best thing to accept, even without understanding.

Acceptance without understanding, trusting in faith, and realizing that hope is what makes a truly good life is what our goal in life is really all about. Facing destruction opens our minds to the realities that are part of our lot in life. We are here by grace...we live as a gift from God, therefore, we should be thankful and appreciate all He allows us. When He disciplines us, we need to take this to heart...something is wrong in our life, and we need to change that mistake...that is what life is all about. Correcting the wrongs, making our attitudes right, and forsaking all that doesn't fit God's great plan for us.

Changing our minds at the last minute isn't a plan...when you understand the words of God, live for Him, and do not look back.

Embrace Him, love Him and do what He wants us to do. Give up selfishness, become humble, and lose your life, for those that lose their life on this Earth will gain life...that is what Jesus meant.

When we stop greed, self promotion, and senseless sin, it's time to realize you are offering God your life...you are willing to die for him, therefore you live for Him. That is the mystery you are trying to figure out.

Lose your life and give it to God...you are really gaining your life...that is the juxtaposition...the ultimate irony.

Jesus said, there are glories and wonders we cannot even imagine...wonders that will be for the righteous, the people who are going to live with Him.

These wonders are beyond what we as humans are able to imagine. We were told that by someone who knows...Jesus. Why are we then so tempted by the sinful things on this Earth, things that we are told are here today and gone tomorrow, are meaningless, and only morally destructive to our inner soul.

Our minds are so limited by what we feel, by what we learn…our brains are in cohorts with our lust and desires, and we stop looking beyond those simple things, when we should realize the most important book on Earth is the Bible, and understanding that book is the most important task we have. These are the words of life…the apostles said to Jesus "Lord, you hold the words of life." These words are so important we sometimes are overwhelmed, taken aback, and shy away, hiding in simple and foolish pursuits.

I realize the Bible holds the key to life, the word of our God, and that nothing matter more than what that book tells us. The wisest human might very well be Solomon, for he was granted wisdom by God…he might not have given us $E=MC^2$, but he gave us timeless wisdom on how to live, and that life is the same everywhere, and that there is nothing new under the sun.

Our great scientific strides might puff up some people, thinking they are almost God-like in their understanding, but that is mere vanity, and that behaviour is exposed by the words of life, and proves that they are meaningless when compared to the word of our God.

Day 48:
This is the end.

Fate

Note to self:

The moment you accept your fate, your situation, your addictions, you loves and hates, your dreams, you tend to dry up creatively and cease to produce. When you give up on your dreams and realize a drab existence, you are immediately doomed to that existence for the rest of your life. Death is part of life, so leave something good behind…it might give you some immortality. Only a righteous person will reap rewards, so keep in mind what you'd like to known for…no one puts up a statue to honor Scrooge or Hitler.

What will be known about you when you die?

Your kids will be aware that they had a different father, but he disappeared.

The people you know here will note your passing, and perhaps be sad, as that is the polite thing to do.

Where I go depend on my faith, so it will not be a tragedy, it will be joyous for me. Love and acceptance…in an unlimited supply. A world of peace and tranquility…serenity and goodness.

How do you get that on earth?

Figure that out, and you'd have heaven on earth…isn't that what that book refers to? I'm in heaven…I'm in heaven. If you live without self-imposed burdens, you see joy.

That was from being happy with yourself…making peace with those around you, and not letting anything bother you. Be strong in your joy…so strong no one can disturb that joy, or spoil your happiness.

Breaking the agreements you made. Like an addict never changes…once a junkie always a junkie…how do you break that piece of self absorbed nonsense?

Tell yourself what you can be, and try something new.

Shed your baggage…life runs out, and yours is just being wasted away, watching T.V., writing stuff no one will probably ever read…it's just a bunch of data…erased in a flash.

It could never have the same existence as a hand written book, in essence, it's too small to be noticed in the grand scheme of things, and you are too apathetic right now…pick yourself off, dust off the years of waste, and live the rest with zeal and love.

Just because you haven't achieved your goals, your early hopes, doesn't mean you can't use something to alter your mundane existence and produce something you can be proud of. We're all good at something, just find out what suits you, and be the best you can be. People are sometimes brainwashed during childhood. We trust adults: so when an adult says something, even when in anger, a child believes what is said…positive and negative. We must learn to break these early agreements we made as a child. If an adult tells you that you can't do something, just go ahead, and try it…you might be the best in the world. But, you have to try…and in order to try, these early agreements you thought we true as a chic turn out to be the mistaken ratings of some idiot that couldn't do what you can do, so decided that you shouldn't do something they couldn't do. There are four major agreements we make as a child, and these must be broken and pushed aside, so you can live your life to the fullest and be all that you can be.

Parents often tell their children they have no talent for things they have no talent for themselves. A selfish attitude, but it happens all the time. Many parents are loving and encourage their children to try everything, supporting them fully in whatever they choose.

Others make assumptions, such as: you can't sing, you can't dance, you have no musical talent, or you are not an artist. The child grows up believing what they were told, until a teacher or other adult encourages them to try their hand at something, and presto, they are sometimes full of talent. Parents should always encourage their children; embrace whatever they try, as long as it doesn't get them into trouble.

Give your brain a chance. You never have. It has been something or another. Don't like your situation? Change it. Empower yourself. Scare the hell out of you by at least thinking it is remotely possible you might be able to do something and succeed. You think therefore you are able. Or, it was only a dream that you thought you could think. You either think or do not think. That is your choice.

If you choose to hope and keep fighting, all bets are off; you may actually earn, learn or win something more. A bitter existence is a chore, a better existence is not beyond you while you can still dream, so do something and change your life.

Being bitter from what you didn't do or what you did do that wasn't right is a waste of a life.

Admit your mistakes, then try, try, try, and try…add lots of try's here, try to do better.

Without darkness, how would you understand light?
To know addiction, you must also know freedom?

To be free, you must know captivity…you know captivity, and what it did as a consequence for breaking the law…

When a child can look like a man, does that mean he knows what it's like to be a grown man?

Maturity is the controlled reasoning of your mental abilities…to show discipline over your urges and desires…

X is a desire…if you learn to exist without X, does that mean you've conquered X, or to understand X, must you also know Y, and therefore understand that both of them together results in Z…therefore knowing X and Y determines you also understand Z, which is the knowledge of both of them. What is the attraction that holds matter together? Is it part of the same force that holds the moon in orbit? Good old gravity? Does any mass, no matter how small, create a force that attracts and forces objects to endlessly circle them? What started this motion? Something like the big bang theory that we say started our universe?

Was it the hand or spark of God, touching a super dense piece of matter that ultimately spread out into our universe? Only God knows these answers, such as where that super dense chunk of every element that creates life come from…how can a scientist ever understand that answer? He may guess about the process…the chain of events that occurred after that initial spark, but showing where that matter came from is beyond science, and in the hands of God.

With such a sure example of the Divine handiwork, why are we not on our knees all the time, praising God, helping others, and doing what He suggests we do with the life He gave us?

Does evil interfere with our free choice? Or, are we unable to resist temptation, fall into sin, which is merely a part of life we must live with, after we learned about it in the Garden of Eden?

These questions, and others, will never be answered by anyone on Earth…the only person we can trust in this is Jesus, who left and returned to Heaven, bringing us the Word of the Father.

Will the world end on December 21, 2012? God said the end is to be better than the beginning, so an end is inevitable, talked about in Revelation, and talked about by Jesus, our Saviour and high priest of humanity. God does not lie, and would never talk about such an end to civilization unless it was a fact. When God says something is a fact, we can take that to the bank. What should we do? Acceptance of our fate is in accord with what our creator has in mind, and anything going against what our creator wishes is going to fail. He made us, therefore He can do what He wills with us. The mold does not ask the mold maker to change what His plans are, that is entirely up to Him.

Building a super-spacecraft is almost a disaster waiting to happen. Unless God allows something to succeed, it will fail, and fail miserably.

There is no way to cheat fate. We consider our freedom to be ours, that there is an alternative life to the one God tells us about in the Bible, but if the Bible is the word and plan of our creator, we can only submit to His will, and trust in His judgment. This is not open to debate. Does this mean we should embrace our death? God did say the end should exceed the start, so thinking along this line is a good idea. We cannot escape death, and only live with what we are given.

We should pray, ask for forgiveness, and turn from sin. That, in the long run, is the only way we are able to ensure eternal life. Trust in God, and do exactly what He says what we need to do, and that is erase sin from our lives, for He is Holy, therefore we must be Holy, and try to achieve this as our main goal in life. Simply put, it is the only goal in life, and we must fight Satan, accept goodness, and say NO to sin and what might go along with that.

Living life with your main goal driving you, do not question it, accept it, and trust that it is for the best. Thinking we are missing something is vanity, pride, and sin. Be humble, be good, and worship the Lord. He knows what is the best for us, and we must embrace that with open arms, and a glad heart. A free heart is an open heart, and we must understand our place in the world…be willing, humble, and full of the gift from God. That is the best thing we could achieve, and the only thing that really matters. Accept this, and you will learn to love life, lose fear, and embrace whatever God has planned for you. Trusting in God is beyond our understanding, but it is the best thing to accept, even without understanding.

Acceptance without understanding, trusting in faith, and realizing that hope is what makes a truly good life is what our goal in life is really all about. Facing destruction opens our minds to the realities that are part of our lot in life. We are here by grace...we live as a gift from God, therefore, we should be thankful and appreciate all He allows us. When He disciplines us, we need to take this to heart...something is wrong in our life, and we need to change that mistake...that is what life is all about. Correcting the wrongs, righting our attitudes, and forsaking all that doesn't fit God's great plan for us.

Changing our minds at the last minute isn't a plan...when you understand the words of God, live for Him, and do not look back.

Embrace Him, love Him and do what He wants us to do. Give up selfishness, become humble, and lose your life, for those that lose their life gain it, which is really what Jesus meant.

When we stop greed, self promotion, and senseless sin, it's time to realize you are offering God your life...you are willing to die for him, therefore you live for Him. That is the mystery you are trying to figure out.

Lose your life and give it to God...you are really gaining your life...that is the juxtaposition...the ultimate irony.

Jesus said, there are glories and wonders we cannot even imagine...wonders that will be for the righteous, the people who are going to live with Him.

These wonders are beyond what we as humans are able to imagine. We were told that by someone who knows...Jesus. Why are we then so tempted by the sinful things on this Earth, things that we are told are here today and gone tomorrow, are meaningless, and only morally destructive to our inner soul.

Our minds are so limited by what we feel, by what we learn...our brains are in cohorts with our lust and desires, and we stop looking beyond those simple

Day 49:

I suppose these are dated. We're still here, 2014.

I would have to say one of the biggest questions right now, is whether we learned from UFO crashes, and have actually built vehicles that are more advanced than anything we can imagine?

Bob Lazar, a whistle-blower that came forward in 1989, said we were working with something called UnunPentium 115. No one knew what he was talking about, but in 2003, several new elements were added to the table of elements, including UnunPentium, and other similar heavy particles. Scary stuff. Did a UFO crash in Germany in 1936, and did they use that technology to try and create an anti-gravity device? Supposedly valid questions.

I have my own UFO experience that makes me take this a bit more seriously than the average person. Is NASA really a cover for the U.S. real space program...something that is in keeping with known technology, but utilizes some of the secrets they've unlocked from crashed UFO sites. When a question has no answer, I am disturbed, but only if it is a valid question that should have answers. I'm not a conspiracy buff, but I think many things are hidden from us, and we are entering into a new frontier that will be mind-boggling...a time when all will be known.

I have enough proof to answer my questions, but I don't have the smoking gun required to prove to the world what I already know.

Is the Iranian's Dr. Keshe able to do what he has claimed? Creating a tractor beam? Wow. However, Iran did bring down a U.S. predator drone without a single scratch. I'm sure they're not too skeptical, and are quite P.O.'ed.

If these fundamental religious people are confronted with life from another world, what would that do to their true religious beliefs? I doubt the Iranian military shared this with the overall population, but it must make their leaders start to question everything they've been taught.

The Bible talks about Heavenly technology, something so dangerous, angel "fell from Heaven," and where did they go? Some other planet? Perhaps things, when we should realize the most important book on Earth is the Bible, and understanding that book is the most important task we have. These are the words of life...the apostles said to Jesus "Lord, you hold the words of life." These words are so important we sometimes are overwhelmed, taken aback, and shy away, hiding in simple and foolish pursuits.

I realize the Bible holds the key to life, the word of our God, and that nothing matter more than what that book tells us. The wisest human might very well be Solomon, for what is the attraction that holds matter together? Is it part of the same force that holds the moon in orbit? Good old gravity? Does any mass, no matter how small, create a force that attracts and forces objects to endlessly circle them? What started this motion? Something like the big bang theory that we say started our universe?

Was it the hand or spark of God, touching a super dense piece of matter that ultimately spread out into our universe? Only God knows these answers, such as where that super dense chunk of every element that creates life come from...how can a scientist ever understand that answer? He may guess about the process...the chain of events that occurred after that initial spark, but showing where that matter came from is beyond science, and in the hands of God.

With such a sure example of the Divine handiwork, why are we not on our knees all the time, praising God, helping others, and doing what He suggests we do with the life He gave us?

Does evil interfere with our free choice? Or, are we unable to resist temptation, fall into sin, which is merely a part of life we must live with, after we learned about it in the Garden of Eden?

These questions, and others, will never be answered by anyone on Earth...the only person we can trust in this is Jesus, who left and returned to Heaven, bringing us the Word of the Father.

www.ingramcontent.com/pod-product-compliance
Lightning Source LLC
Chambersburg PA
CBHW072047190526
45165CB00019B/1998

that's where all these ET's are from, and what they really are. The worst case scenario might have them be demons...pure evil entities that love death and destruction. If these demons were working with the U.S. and giving them technology, death will follow.

I hope the religious questions I dealt with could help someone with a similar attitude. Reaching out and helping someone find their spiritual path is part of the Christian routine, but standing on a street corner, handing out pamphlets, doesn't work. Standing on a soap box will land me in the nut ward, so this is the next best method of getting the message out there. Pick up your Bible and learn the truth. It's not trendy or high-tech, but it restores you soul, and is healing to your bones. Don't be "too wise in your own eyes" and consider yourself about simple Christian doctrine. There may be many forms of it, but the basic premise is sound, true, and from the omnipotent creator of the universe. Just before the big bang, when that super-condensed ball of matter existed, I prefer to think the finger of God touched it off, rather than some meaningless scientific action. I see intelligent design everywhere; God is a mater scientist and knows all things, I'd bank on that anytime. Picturing a cold line of evolution that eventually produced us is a drab and lifeless approach to a universe teeming with unknown lifeforms. We can probe the depths of our souls, and that is something no scientist can put in an equation, or a test tube. Just believe, and pursue life.

Selah